A Case
Against Accident
and
Self-Organization

A Case
Against Accident
and
Self-Organization

Dean L. Overman

ROWMAN & LITTLEFIELD PUBLISHERS, INC.
Lanham • Boulder • New York • Oxford

Quotation reprinted with the permission of Simon & Schuster
from *Catch-22* by Joseph Heller. Copyright © 1955, 1961 by
Joseph Heller. Copyright renewed (c) 1989 by Joseph Heller.

Quotations reprinted with the permission of Adler & Adler from
Evolution: A Theory in Crisis by Michael Denton. Copyright ©
1985 by Michael Denton.

Quotations reprinted with the permission of Cambridge Univer-
sity Press from *Information Theory and Molecular Biology* by
Hubert Yockey. Copyright © 1992 by
Cambridge University Press.

Library of Congress Cataloging-in-Publication Data

Overman, Dean L.
    A case against accident and self-organization / Dean L.
Overman.
        p. cm.
    Includes bibliographical references and index.
    ISBN 0-8476-8966-2 (cloth : alk. paper)
    1. Life—Origin. 2. Molecular biology. 3. Probabilities.
4. Self-organizing systems. 5. Cosmology. 6. Nuclear
astrophysics. 7. Evolution—Philosophy. I. Title
QH325.O84   1997
576.8'3—dc21   97–25885
                                                    CIP
        ISBN 0-8476-8966-2 (cloth: alk. ppr.)

*This book is dedicated to Linda,
Christiana, and Elisabeth.*

CONTENTS

FOREWORD

This book argues persuasively against the assumption that the origin of life and the origin of the universe can be accounted for as random events. According to Overman, it is mathematically not possible to derive the origin of the high level of information necessary for organic life in terms of random fluctuations in pre-organic processes. The author does not exclude the idea of self-organization as the cause of the origin of life by *a priori* reasons, but he offers sobering arguments against contemporary hypotheses of self-organization. The arguments are presented in a very lucid and informative manner and provide useful reading for theists and for naturalists. The point of Overman's argument against accident is not to disregard the importance of contingency in natural events and processes, but to consider events like the origin of life or the origin of the universe as part of a coherent system that has to be accounted for in its total wholeness. This concern becomes even more apparent in the discussion of modern cosmology. In this discussion, the author not only provides a very well written and detailed survey of the present situation in cosmological debate, but also convincingly argues against positions that try to avoid the consequences of the Big Bang model of cosmology which suggests a unique history for our universe. The section on speculative alternatives to the standard theory of astrophysics is of special interest in this respect, particularly the detailed arguments against the ideas of Stephen Hawking and the discussion on the limited value of the anthropic principles. The theological concerns of the author are not used as arguments. The author is well aware of the fact that they belong to another level of discussion. But he provides the basis for such a discussion with his thesis that the origin of the universe has to be accounted for as a coherent whole and the requirement of a cause of the being of the universe cannot be evaded by assumptions concerning an accidental beginning *in fieri*. The detailed and lucid argument for his thesis constitutes a valuable contribution to the discussion about the relationship between natural science and a theology of nature.

Professor Dr. Wolfhart Pannenberg
University of Munich

ACKNOWLEDGEMENTS

This book depended upon the assistance and good will of many persons. Lana Couchenour, my colleague and assistant for over twenty years, gave me her balanced sense of judgment and her consistent capacity for tact and truth. I am grateful for my appointment as a Visiting Scholar and Officer of Harvard University and for the use of the fine facilities of that institution; the staff of the Harvard Faculty Club always made my stays in Cambridge pleasant and extended their courtesy to my daughter, Elisabeth, an undergraduate at the time at Harvard College. My daughter, Christiana, contributed her considerable skills in geometric reasoning and drawing to assist in the understanding of some abstract concepts in physics. I am also grateful for the encouragement and comments of Dr. Michael Behe, a biochemist at Lehigh University, Dr. Robert Kaita, a particle physicist at Princeton University, Dr. Alister McGrath, Principal of Wycliffe Hall, Oxford University, who holds a doctorate in molecular biology and a degree in theology, Dr. Armand Nicholi, professor at the Harvard Medical School and Harvard College, and Dr. John Polkinghorne, formerly professor of mathematical physics at Cambridge University. I am also indebted to Dr. Hubert Yockey, a brilliant physicist who studied under J. Robert Oppenheimer at the University of California at Berkeley and later worked with him on the Manhattan Project. Much of my thinking about complexity was influenced by his writings and by my conversations and meetings with Dr. Yockey, who is the preeminent authority on information theory and molecular biology.

My wife, Linda, maintained great patience and support as the writing of the manuscript occupied much of my time during a vacation in Macatawa, Michigan. Macatawa is the beachfront cottage community where Frank Baum wrote *The Wizard of Oz* at the beginning of the twentieth century. Walking on that beach at the end of the twentieth century and reflecting on the bizarre concepts of extra dimensional string theory which allow components of quarks to turn into black holes and vice versa, I emphathized with Dorothy's experience and her attempt to return to reality from a very strange world.

PREFACE

I never intended to write this book. I have no desire to enter the debate concerning the origin of life on earth. I am a lawyer concerned with logic and the validity of premises, inferences and conclusions as they relate to an examination of evidence. This book is simply the result of my reading an article in *Telicom*, the journal of the International Society for Philosophical Enquiry (ISPE). ISPE is an organization consisting of approximately 750 members from more than 40 countries who have tested at the 99.9 percentile on certain standardized reasoning tests and who carry on a variety of discussions through *Telicom* or by personal correspondence. The article in *Telicom* set forth the proposition that because the Miller and Urey experiment "works," individuals are free to select their own purposes and goals without regard to any standard. My first response was to write a draft of a five page letter to the editor explaining my objections to certain propositions in the article. I then noticed in the directions for the submission of materials that the journal prefers letters to the editor with a maximum length of two pages. Because I could not condense my arguments to two pages, I expanded the letter to the present length of this book. Recalling the old apology, "I am sorry that I wrote you such a long letter; I didn't have the time to write you a short one," I hope I have not obfuscated my argument by increasing the verbiage.

Dean L. Overman
Washington, D.C.

POWERS OF TEN

In reviewing mathematical probabilities in molecular biology and particle astrophysics, we will sometimes work with very large and very small numbers. For clarity, a power of 10 is used. A power of ten is simply a number with a base of 10 and a logarithm or exponent given to the base. When the number 10 is used as a base, the exponents are known as common logarithms. 10^n is equal to 1 followed by n zeros. 10^4 is a shorthand for 10 x 10 x 10 x 10 or 10,000. 10^{-4} is equal to $10 \div 10 \div 10 \div 10$ or .0001. Thus 10^6 = one million, 10^9 = one billion, 10^{-6} = one millionth, and 10^{-9} = one billionth.[1] The following table may be useful:

10^{10} =	10,000,000,000
10^9 =	1,000,000,000
10^8 =	100,000,000
10^7 =	10,000,000
10^6 =	1,000,000
10^5 =	100,000
10^4 =	10,000
10^3 =	1,000
10^2 =	100
10^1 =	10
10^0 =	1
10^{-1} =	0.1
10^{-2} =	0.01
10^{-3} =	0.001
10^{-4} =	0.0001
10^{-5} =	0.00001
10^{-6} =	0.000001
10^{-7} =	0.0000001
10^{-8} =	0.00000001
10^{-9} =	0.000000001
10^{-10} =	0.0000000001

We can also represent a number by multiplying a number times a base of 10 with a common logarithm. For example, 3×10^6 = 3 million. 2.5×10^6 = 2.5 million. 2.5×10^3 = 2,500. 3×10^{-6} = 3 divided by one million or 3 millionths.

Using an exponential system for very large or very small numbers is useful, because one can see immediately that 10^{50} is a larger number than 10^{49}; but this is not so obvious in comparing 100,000,000,000,000,000,000,000,000,000,000,000,000,000,000,000,000 to 10,000,000,000,000,000,000,000,000,000,000,000,000,000,000,000,000.

Most mathematicians consider a probability of less than one in 10^{50} as mathematical impossibility.

PART I

INTRODUCTION

This book reviews the evidence from discoveries in molecular biology and physics in the context of mathematical probability calculations and considers the question of accident in the formation of a universe compossible with living matter and the formation of the first living matter. The book also analyzes the plausibility of self-organization scenarios in the formation of the first living matter. The first question presented in this book is not whether evolutionary processes occurred; but whether the processes were accidental and by chance. Specifically, the first question presented is: under standard probability definitions, is it mathematically possible that accidental or chance processes caused (a) the formation of the first form of living matter from non-living matter and (b) the formation of a universe compossible with life? The second question presented is: are current self-organization scenarios for the formation of the first living matter plausible?

Even within the boundaries of mathematical possibility, an objective, reasonable person following the principles of the scientific method will favor a proposition which has a probability of .999 over a proposition which has a probability of .001. Metaphysical predilections, however, can impede a person's scientific objectivity and cause him or her to select the low probability proposition. Many otherwise rational persons make unwarranted conclusions which are not based on evidence, but are made in the absence of evidence and contrary to mathematical probabilities because of their faith in the ideology of materialism. This book evaluates those conclusions from a quantitative perspective. Because mathematicians normally regard anything with a probability of less than one in 10^{50} as mathematical impossibility, we will use that standard in answering the first question presented. For the purpose of making the book more readable to a general audience, most of the supporting mathematical calculations are contained in the endnotes to the book.

This book begins with a review of the influence of metaphysical assumptions in logical analysis and the issues raised in the use of logic under a presupposition that thought is a product of accident. After a discussion of some principles of logic applicable to the questions presented and the limitations of logic, I present a definition of life, discuss the genetic code, and review the theory of the emergence of life from accidental or chance processes, including the experiments allegedly supporting that theory. I then examine the evidence for the prebiotic soup which is the foundation for the theory and analyze the time available for the formation of life on earth. I review the calculations of mathematical probabilities of abiogenesis[2] from chance processes and discuss the need of proponents of self-organization scenarios to identify a mechanism for generating sufficient information content into inert matter within the context of the definition of life. I consider the implications of ALH84001, a meteorite containing possible evidence of remnants of life on Mars. With an understanding of the distinction between living and non-living matter and the role of complexity or information content in that distinction, I then discuss the precision of values in certain aspects of particle astrophysics necessary for life and the explanations offered by the weak and strong anthropic principles. Finally, I discuss the difficulty in forming an adequate foundation for ethics with any world view consistent with accident or any other impersonal cause for the formation of the universe and the first living matter.

PART II

VERBAL AND MATHEMATICAL LOGIC RELATING TO THE QUESTIONS PRESENTED

2.1. Influence of metaphysical assumptions

Complete objectivity in science is an illusion. Because so much of one's analysis depends upon metaphysical assumptions, it should be acknowledged by this writer, and by all readers, that the answer one gives to a question depends to a great extent on the metaphysical position one has previously adopted.[3] This is certainly true for theists, and it is equally true for materialists. Frequently, the metaphysical conclusion is given as the rationale for a tortured interpretation of evidence. Theists and naturalists frequently refuse to follow evidence where it leads on the basis that to do so would result in a contradiction of their previous metaphysical conclusions. *Quod volumus, facile credimus.*[4]

No one is immune from making mistakes because of his or her metaphysical assumptions. Even Albert Einstein said that the biggest blunder of his life occurred when he allowed his own world view to force him into creating a "cosmological fudge factor" to keep his mathematical formula consistent with a steady state and infinite universe. Einstein's equations in his general theory of relativity actually predicted an expanding universe. To avoid this conclusion, he put an extra term into the equations which cancelled out the expansion. George Gamow, who worked under Professor Alexander Friedman, a Russian astronomer who first noticed Einstein's mathematical error, described the error and Einstein's discussions with him:

> It is well known to students of high-school algebra that it is permissible to divide both sides of an equation by any quantity, provided that this quantity is not zero. However, in the course of his proof, Einstein had divided both sides of one of his intermediate

3

equations by a complicated expression which, in certain circumstances, could become zero. In the case, however, when this expression becomes equal to zero, Einstein's proof does not hold, and Friedman realized that this opened an entire new world of time-dependent universes: expanding, collapsing, and pulsating ones. Thus Einstein's original gravity equation was correct, and changing it was a mistake. Much later, when I was discussing cosmological problems with Einstein, he remarked that the introduction of the cosmological term was the biggest "blunder" he ever made in his life.[5]

2.2. Thoughts as products of accidents

Even assuming that one can minimize his or her metaphysical presuppositions in reviewing evidence, another assumption must be made before we can examine the theory of the emergence of life from unguided, chance processes. The proposition that a universe compossible with life and the first form of life developed by accident rather than by design raises the following conundrum: if logical thinking is an accident, is it trustworthy? Or, to modify the enigma, is it probable that accidents will accurately describe other previous accidents? The concept that the universe and our existence were the products of accidents means that all our thinking is merely the accidental result of accidents. But if your thoughts and my thoughts are only accidents (are not results of accidents also accidents themselves?), then why should you or I consider our thinking true or logical? Isn't it only accidental? How can we trust thought if it is an accident?

If we rely on mathematical and verbal logic in evaluating the question of the emergence of life by accidental or unguided, chance processes, we face the difficulty of assuming the answer to the question prior to the evaluation of the evidence. If mathematical and verbal logical thought processes are our method, however, we must begin with the assumption that, whether or not our thoughts are the products of accidents, mathematical and verbal logical processes are valid. In order to evaluate the evidence presented in this book from the perspective of a materialist and from other perspectives, despite the problem stated above, we must hold in abeyance any conclusion that thinking cannot be logical if it is the product of accidents.

2.3. Valid and false reasoning

As with any subject matter setting forth propositions and theories, some of the relevant literature involving molecular biology and particle astrophysics have fallacies in reasoning which lead to illogical and invalid conclusions. To assist in our efforts to maintain sound forms of analysis, we will discuss some of the forms of errors in reasoning which appear from time to time in the literature. We will also discuss limits on logic, including Gödel's Incompleteness Theorem in mathematics and the Heisenberg uncertainty principle in quantum mechanics. Finally, we will raise the question of the decidability of (a) any origin of life scenario which requires the explanation of a method of generating sufficient information content into inert matter to qualify such matter as living under the definition given in this book and (b) any theory of physics which requires a knowledge of events "prior" to Planck time (10^{-43} of the first second or time zero).

2.3.1. Valid and invalid syllogisms

For purposes of this book, structured reasoning or logic is assumed to be valid. As indicated above, if we do not assume that we can trust logic, we can proceed no further. We are without valid means to examine the evidence. Let us begin with a discussion of valid forms of syllogisms. The distinction between valid and invalid syllogisms is elementary, but a brief review of a few examples is worthwhile. A syllogism is valid in form where one can draw an inexorable conclusion from the premises and facts. *Ab universali ad particulare valet.*[6] An example of a valid syllogism is:

> All golden retrievers are dogs.
> Toby is a golden retriever.
> Therefore Toby is a dog.

Some syllogisms, however, can appear correct but have unwarranted conclusions:

All pit bulls are dogs.
Some of the smartest animals are dogs.
Therefore some of the smartest animals are pit bulls.

In this invalid syllogism, the conclusion does not allow for the implied possibility that pit bulls may not be among the smartest dogs. The illogical nature of this type of invalid syllogism is more dramatic in the example: All condors eat carrion. Some sharks eat carrion. Therefore some condors are sharks. One cannot move from a "some" statement to a conclusion that depends on the initial "all" assumption.

To review an example more pertinent to this book, consider the following invalid syllogism with an unwarranted inference: The amino acids were produced in the mixture contained in the glass apparatus after an electrical charge was sent into the mixture. Most of the mixture is like the early earth's atmosphere which probably had additional ingredients. Therefore, the amino acids were produced in a similar way in the early earth's atmosphere.

This is an invalid syllogism, because most is not all, and there are ingredients in the early earth's atmosphere not included in the mixture so the mixture and the atmosphere are not the same. Again, one cannot move from a "some" statement to a conclusion that depends on an "all" premise. The two are not identical and do not demand an inexorable conclusion. *A particulari ad universale non valet consequentia.*[7]

A good evaluation of reasoning requires that one examine both the reliability of the inferences and the reliability of the premises. Acceptable reasoning also requires an inexorable conclusion from valid inferences from the premises. For example, in reviewing the phrase, "complexity on the edge of chaos," the reader should question the precise meaning of the terms and the validity of the inference from the evidence to see if the statement is inexorable from the premises or only a play on words such as that found in the following exchange from *Alice in Wonderland*:

The Hatter opened his eyes very wide on hearing this; but all he *said* was "Why is a raven like a writing-desk?"

"Come, we shall have some fun now!" thought Alice.

"I'm glad they've begun asking riddles—I believe I can guess that," she added aloud.

"Do you mean that you think you can find out the answer to it?" said the March Hare.

"Exactly so," said Alice.

"Then you should say what you mean," the March Hare went on.

"I do," Alice hastily replied; "at least—at least I mean what I say—that's the same thing, you know."

"Not the same thing a bit!" said the Hatter. "Why, you might just as well say that I see what I eat is the same thing as I eat what I see !"

"You might just as well say," added the March Hare, "that I like what I get is the same thing as I get what I like !"

"You might just as well say," added the Dormouse, which seemed to be talking in its sleep, "that I breathe when I sleep is the same thing as I sleep when I breathe !"

"It *is* the same thing with you," said the Hatter, and here the conversation dropped, and the party sat silent for a minute, while Alice thought over all she could remember about ravens and writing-desks, which wasn't much.[8]

2.3.2. *Extrapolations from a small amount of data*

Logical fallacies include extrapolations from a small amount of evidence. Extrapolations from a small amount of data often result in unwarranted inferences. One must be careful in making broad generalizations from a single observation of a small amount of data. Extrapolation fallacies are common in the literature containing speculations about the origin of life. There is a vast difference between a theory like the theory of general relativity in physics which is supported by empirical verification of mathematical predictions and the theory of chance emergence of life based on speculations from extrapolations of a small amount of questionable data produced by the Miller and Urey line of experiments. Mark Twain was well aware of extrapolation fallacies when in *Life on the Mississippi* he made the following sardonic calculations of the past and future length of the Mississippi River based on an unwarranted inference that the Mississippi could continue to shorten at a constant rate:

> Therefore the Mississippi . . . was twelve hundred and fifteen miles long, one hundred and seventy-six years ago. It was eleven hundred and eight after the cut-off of 1722. It was one thousand and forty after the American Bend cut-off. It has lost sixty-seven miles since. Consequently, its length is only nine hundred and seventy-three miles at present.
>
> Now, if I wanted to be one of those ponderous scientific people, and "let on" to prove what had occurred in the remote past by what had occurred in a given time in the recent past, or what will occur in the far future by what has occurred in late years, what an opportunity is here! Geology never had such a chance, nor such exact data to argue from! Nor "development of species," either! Glacial epochs are great things, but they are vague—vague. Please observe:
>
> In the space of one hundred and seventy-six years the Lower Mississippi has shortened itself two hundred and forty-two miles. That is an average of a trifle over one mile and a third per year. Therefore, any calm person, who is not blind or idiotic, can see that in the Old Oölitic Silurian Period, just a million years ago next November, the Lower Mississippi River was upward of one million three hundred thousand miles long, and struck out over the Gulf of Mexico like a fishing rod. And by the same token any person can see that seven hundred and forty-two years from now the Lower Mississippi will be only a mile and three-quarters long, and Cairo and New Orleans will

have joined their streets together, and be plodding comfortably along under a single mayor and a mutual board of aldermen. There is something fascinating about science. One gets such wholesome returns of conjecture out of such a trifling investment of fact.[9]

2.3.3. *Inconsistencies within the context of terms*

Inconsistencies within the context of the terms of the issues presented often result in invalid reasoning, especially where distinctions are not drawn between incomparable terms. In later sections we will distinguish between order and complexity, terms which are frequently confused in origin of life scenarios. For the moment, consider the following example: assume that a driver wants to drive a car a distance of two miles at an average rate of 60 miles per hour. Because of traffic, the driver discovers that at the end of the first mile the car has only averaged 30 miles per hour. What average speed does the driver need to drive the second mile to complete the goal of averaging 60 miles per hour for the two miles?

The answer is not 90 miles per hour, but something faster than the speed of light (186,262 miles per second). The analysis of average rates must be done over the terms for equal time intervals; the terms for equal distance intervals are not relevant (just as order is not as relevant as complexity in theories concerning the formation of the first living matter). From a time perspective, the first mile was driven in two minutes (30 miles per hour x 2 minutes = 1 mile). To average 60 miles per hour for two miles, however, the driver must drive the two miles in two minutes. But the driver has already used two minutes to drive only the first mile. The driver has no more time allowed for the second mile.

Analyzing the question from a perspective of time intervals allows for consistencies in context. If the driver drove 30 miles per hour for only one minute, and then 90 miles per hour for the second minute, the driver would travel the two miles in two minutes and thus average 60 miles per hour. The solution to the problem requires an analysis of time intervals, not distance intervals.

Consistent use of terms is essential for valid reasoning. Because the commingling of terms is so prevalent, I assume the risk of being redundant by emphasizing that theorists have confused the concepts of order and complexity. Many authors use the terms "order" and "complexity" as synonyms or with subtle changes in meaning in the term "complexity" which appear to strengthen an argument. As a result, many theories make questionable analogies from the generation of order out of chaos in the inorganic world to the generation of complexity in the organic world. As we shall discuss, the terms "order" and "complexity" can be opposites; order has very little to do with complexity as that term is defined in this book. As George Johnson has written in his recent book, *Fire in the Mind*, "complexity can be a maddening slippery concept."[10] This is especially true when a variety of meanings for the term are used by a theorist even in the same book. (We will later discuss that theorists at the Santa Fe Institute use more than thirty different definitions of complexity.) In this book I am using complexity univocally with the precise definition given by information theory, where complexity relates to the level of information content in a structure. To allow such a key word to change its meaning in the course of a description of a theory commits the logical fallacy of equivocation. When the change in meaning is subtle, an unwarranted conclusion may appear to follow validly from the premises. Consider the following controversial example:

> Only man is rational.
> No woman is a man.
> Therefore, no woman is rational.

This argument, not only is unwise because of the imminent danger any male using it would encounter, but also invalid because the word *man* has a different meaning in the second sentence than it does in the first sentence. In the first sentence, man means *human*; in the second sentence, man means *male*. Equivocation can make some arguments appear to be sound when actually they are only a play on words.[11]

The fallacy of equivocation can occur any time one uses inconsistent meanings for a word. Like complexity, many words have several meanings. Literature is replete with fallacious reasoning from equivocation. Lewis Carroll, a mathematician, used this fallacy to provide some humor in *Through the Looking Glass*:

"Who did you pass on the road," the King went on, holding out his hand to the Messenger for some hay.
"Nobody," said the Messenger.
"Quite right," said the King, "this lady saw him too. So of course Nobody walks slower than you."
"I do my best," the Messenger said in a sullen tone.
"I'm sure nobody walks much faster than I do!"
"He can't do that," said the King, "or else he'd been here first."[12]

When the fallacy of equivocation is used by a theorist in describing complexity and order emerging from chaos, the theory takes on plausibility, but not within the definition of complexity used by information theory. An example of the latter definition of complexity is the genetic code which contains a very high level of information content. Living systems have irregular, aperiodic, not ordered, information content. The following example contains the fallacy of equivocation in the use of the terms "order" and "complexity."

The pulling of a drain plug allows the force of gravity to move water from a chaotic, random equilibrium to an ordered vortex form.

Order occurs spontaneously in a system far from equilibrium.

The complexity of living matter can be formed spontaneously in a system where an energy force moves that system far from equilibrium.

The third sentence does not follow from the first two sentences, because order and complexity are different concepts and, in many respects, opposites.

2.3.4. Hidden assumptions and contrivances in mathematics

Mathematics is a wonderful medium for logical reasoning, but only beneficial when the assumptions employed are valid. Even

in strictly mathematical logic, vigilance is required to avoid inconsistencies derived from false or unproven assumptions, which, of
course, lead to inaccurate conclusions. For example, consider the
following algebraic equations:

$$x = y$$
$$x^2 = xy$$
$$x^2 - y^2 = xy - y^2$$
$$(x+y)(x-y) = y(x-y)$$
$$x+y = y$$
$$2y = y$$
$$2 = 1$$

The flaw in the reasoning is hidden in the assumptions. If x is
equal to y, x-y=0. When x-y is cancelled out of each side of the
equation, x+y and y are each divided by x-y. But x-y=0, and
mathematical logic does not allow division by zero. The flaw in
the analysis is a hidden assumption that division by zero will
yield a valid number. As described by George Gamow, *supra*, this
was also the hidden fallacy in Einstein's cosmological fudge factor. Division by zero is impossible because division is defined as
the inverse of multiplication. Fifty divided by five equals a number which when multiplied by five will equal fifty. But fifty divided by zero does not equal a number which when multiplied by
zero will equal fifty. No such number exists. The answer is not infinity, because infinity is not a number.

A more deeply hidden false assumption appears in the following analysis written on a wall in the mathematics building at
Cornell University:

$$(m+1)^2 = m^2 + 2m + 1$$
$$(m+1)^2 - (2m+1) = m^2$$
$$(m+1)^2 - (2m+1) - m(2m+1)$$
$$= m^2 - m(2m+1)$$
$$(m+1)^2 - (m+1)(2m+1) + \tfrac{1}{4}(2m+1)^2$$
$$= m^2 - m(2m+1) + \tfrac{1}{4}(2m+1)^2$$
$$[(m+1) - \tfrac{1}{2}(2m+1)]^2 = [m - \tfrac{1}{2}(2m+1)]^2$$
$$(m+1) - \tfrac{1}{2}(2m+1) = m - \tfrac{1}{2}(2m+1)$$
$$m+1 = m$$
$$1 = 0$$

The false assumption in this analysis appears in the line $[(m+1-\frac{1}{2}(2m+1)]^2 = [m-\frac{1}{2}(2m+1)]^2$. This line means that $[\frac{1}{2}]^2 = [-\frac{1}{2}]^2$. But the square roots of those values are $\frac{1}{2} = -\frac{1}{2}$. This equation, of course, is impossible.[13] The square of a real number is never negative. The equation $x^2=y$ has no real number solution if y < 0. Accordingly, a truly negative number has no real square roots. In this book we will review some unproven assumptions which manipulate mathematical probability calculations. For present purposes we need only note that to make an honest, objective assessment, one must assiduously inspect any theory, whether expressed verbally or mathematically, to uncover unproven or false assumptions.

When I was pursuing graduate studies at the University of Chicago, a school known for its emphasis on quantitative procedures, I noticed that mathematical equations can be used as contrivances to yield a desired result. After working through a long and tedious equation which filled most of a large blackboard, a student questioned whether the minus sign in the last of a series of parentheses should actually be a plus sign. The facetious response was that it didn't matter which sign was used as long as the result was consistent with the answer one wanted.

The absence of false assumptions in mathematical equations is not a complete assurance of valid logic. One should examine whether the equations are mere contrivances or tricks manipulated to yield an answer in support of a metaphysical predilection. Mathematical equations are susceptible to such manipulation and have been used in some very imaginative processes to produce results consistent with a philosophical presupposition. To understand the basic concept of this process, consider the following simple examples where the manipulator structures what appears to be an unbiased formula using arbitrary numbers to produce a desired result.

Assume that you want a friend to perform a mathematical calculation which will always produce the number 7 as an answer. Ask your friend to think of a number, not to tell you the number, and to double the number. Then instruct him or her to add 14 to the number and divide the total by two and then subtract the original number from the quotient. No matter what number your friend selects as the original number, the answer to the mathematical instructions will always be 7, because you have structured a mathematical contrivance which will always produce that result.

The algebra is very simple: your friend selects number N. He or she doubles the number and adds 14 which equals 2N + 14. Dividing 2N + 14 by 2 equals N + 7. Subtracting the original N yields the number 7. The only calculation which is not cancelled out in the instructions or equations is the division of 14 by 2 which, of course, will always yield the number 7.

Consider another simple example of a contrived equation which appears to allow for random changes but will always produce the same answer without regard to the original number selected. Ask your friend to think of any three digit number in which the first digit is larger than the last digit and to reverse that number and subtract it from the original number. Then instruct him or her to add the difference to the number obtained by reversing the digits of the difference, adding 0 as the initial digit if the difference is only a two-digit number. The answer will always be 1089 no matter what three digit number is originally selected.

Again, the algebra is simple: assume that your friend's original number N has as its ordered digits zyx. This number has the following algebraic expression: $100z + 10y + x$. Reversing that number we have $100x + 10y + z$. Subtracting the reversed number from the original equals $100 (z-x) + (x-z)$ which equals $99 (z-x)$. Thus, the answer at this point in our instructions will always be a multiple of 99. Every multiple of 99 (099 when the multiple is one) when added to the reversal of its digits will always result in the number 1089.[14]

These two simple examples illustrate how an equation can be used as a contrivance or trick to produce a predetermined result even when arbitrary numbers are initially selected. Mathematical contrivances designed to produce a predicted number from numbers which appear to be selected arbitrarily have been part of number theory for many centuries. Arranging equations to arrive at a predetermined result may appear more feasible when one examines some of the characteristics of number theory, such as Fibonacci numbers, named after Leonardo de Fibonacci, an Italian mathematician who lived between 1170 and 1240. The Fibonacci sequence begins as follows: 1, 1, 2, 3, 5, 8, 13, 21, . . .

Notice that each number after the first two is the sum of the preceding two numbers,[15] e.g., $5 + 8 = 13$.

The Fibonacci sequence pervades many portions of the physical world. The sequence defines the spirals and counter spirals in the structures of flowers, pine cones, artichokes, leaf stubs on palm trees, broccoli florets, the seeds in a sunflower, and the branches of a bee's family tree. The sequence almost appears to be a mathematical code in nature left by an intelligence. In some respects this code is similar to our sending signals out in the universe with mathematics relating to pi to communicate that intelligence exists on earth in the hope that other extraterrestrial intelligent beings, if they exist, will recognize the mathematics and know that intelligent life exists on earth. In this sense, the Fibonacci sequence can be interpreted as evidence against accident as a cause for these natural structures in the physical world.

The Fibonacci sequence can be very useful in creating contrived numbers. The sequence has predictable characteristics. If after the number 3, one divides any of the Fibonacci numbers by the next higher number, the answer will always be 0.625. If after the number 89, one divides a Fibonacci number by the next highest number, the answer will always be 0.618. With higher numbers, the division fills in more decimal places. If after the number 2, one divides any Fibonacci number by its *preceding* number, the answer will always be 1.6. If after the number 144, one makes such a division, the answer will always be 1.618.[16]

Understanding the predictable characteristics of Fibonacci numbers, one can comprehend how equations using these numbers can be arranged to produce predetermined results. For the sake of simplicity we will demonstrate how Fibonacci numbers can be used to arrive at a contrived number rather than at a contrivance which also produces a predetermined result. In his book, *Liber Abaci*, Fibonacci borrowed an ancient contrivance from the Chinese Remainder Theorem which first appeared about 200 A.D. in Sun Tzu's work, *Sun Tzu Suan-Ching*.[17] To produce the Fibonacci contrived number, ask a friend to select any secret number less than 105. Then ask your friend to divide the secret number by 3, by 5 and by 7 and tell you the remainder from each division. Multiply the first remainder (from the division by 3) by 70; multiply the second remainder (from the division by 5) by 21;

and multiply the third remainder (from the division by 7) by 15. Then add the numbers derived from your multiplication. To the extent the numbers so added exceed 105, subtract 105 to arrive at the contrived number. If the added numbers do not exceed 105, the sum of the added numbers is the contrived number.

To illustrate this contrivance, assume that your friend secretly selects the number 26. Dividing by 3, your friend tells you that the first remainder is 2. You multiply this remainder by 70 with a result of 140. Dividing 26 by 5, your friend then tells you that the second remainder is 1. You multiply 1 by 21 with a result of 21. Dividing 26 by 7, your friend tells you that the third remainder is 5. You multiply 5 by 15 with the result of 75. You then add the results of your multiplication (140 plus 21 plus 75) to determine the sum of 236. The extent to which 236 exceeds 105 is 131. Subtracting 105 from 131, you arrive at the contrived number 26.

To illustrate the contrivance when the addition results in a number less than 105, assume that your friend secretly selects the number 78. Dividing 78 by 3, your friend tells you that the remainder is zero. Dividing 78 by 5, your friend tells you that the remainder is 3. You multiply 3 by 21 with the result of 63. Dividing 78 by 7, your friend tells you that the remainder is 1. You multiply 1 by 15 with the result of 15. The sum of the addition of the results of your multiplication is a number less than 105 so you do not subtract 105. The sum of such addition is the contrived number 78 (zero plus 63 plus 15).[18]

Fibonacci numbers and the other examples given indicate how an equation can be used as a contrivance or trick to produce a predetermined result even when arbitrary numbers are initially selected. Later in the book we will raise the question whether Stephen Hawking's controversial use of imaginary numbers in imaginary time is a more complicated and elegant example of a mathematical contrivance designed to support a metaphysical predilection and avoid a singularity or a beginning boundary to the universe. In an article published in *Science Spectra* in 1996 entitled, "The Human Paradox: Stephen Hawking and His Work," Gordon Fraser notes that Hawking admits that his equations are constructed in such a way as to determine a result which is consistent with no beginning to the universe:

In constructing a quantum mechanical picture of the universe, Hawking says, "we will solve these equations subject to the conditions that the universe has no boundary." In other words, spacetime is completely self-contained, and the equations of the mathematical framework are constructed in such a way as to ensure this. As Hawking says, "There would be no singularities, and the laws of science would hold everywhere, including at the beginning of the universe. The way the universe began would be determined by the laws of science."[19]

Starting with the result one wants and working backwards is not unusual and not always an incorrect procedure. In many circumstances reasoning backwards is a very useful device for constructing plausible hypotheses, but it cannot be substituted for a rigorous proof. It is merely a contrivance to assist one's thinking. Regressive reasoning or reasoning backwards begins with the desired end and then asks what antecedent could produce that end. After determining that antecedent, one asks what could produce that antecedent and so on until *one moves backwards to something already known to be true.* This is a type of heuristic reasoning which serves to discover truth, but only as a device to assist thinking, not as a verification of truth.[20] As we shall discuss, Hawking emphasizes that his no boundary concept is only a proposal and not even a theory. He admits that *it cannot be deduced from another principle already known to be true.* He uses regressive reasoning without any verified principle in the sequence to support his proposal.

2.3.5. Circular reasoning

Another faulty form of reasoning assumes the conclusion in the premises. Note the circular reasoning in the sentence, "You can never expect science to engage philosophy, because science cannot be philosophical." One must examine the premises in any argument to ascertain if the reasoning is *circulus in probando.* Consider the following sentence:

> The primitive atmosphere must have contained re-
> ducing equivalence in some form to yield amino ac-
> ids, since no biomolecules or their precursors are
> formed when a mixture of carbon dioxide, water, and
> nitrogen is sparked.[21]

The reasoning is faulty because it is circular. The conclusion is assumed in the premise. Assumed contradictions are often hidden premises in arguments. For example, the question, "If a designer designed the universe, who designed the designer?", assumes the contradiction by asserting that the designer was designed. Such an assertion is an assumed contradiction hidden in the question. This is similar to asking the question: who or what made triangles circular?

One may argue that if everything has a cause, then a designer must have a cause. Given the assumption in the dependent clause, the conclusion follows logically. If the assumption, however, was modified to: if everything that has a beginning has a cause, the conclusion would not follow if the designer was defined as something that does not have a beginning. If this modification was made and applied to the universe, the argument could be stated:

> Everything that has a beginning has a cause.
> The universe had a beginning.
> The universe must have a cause.

What appears to be an example of circular reasoning was the main theme in Joseph Heller's *Catch-22*. Yossarian, a bombadier, wants to be grounded from combat flying missions. In the following conversation he learns about Catch-22 from Doctor Daneeka:

> "Can't you ground someone who's crazy?"
> "Oh, sure, I have to. There's a rule saying
> I have to ground anyone who's crazy" . . . .
>
> Yossarian looked at him soberly and tried another ap-
> proach. "Is Orr crazy?"
> "He sure is," Doc Daneeka said.
> "Can you ground him?"

"I sure can. But first he has to ask me to.
That's part of the rule."
"Then why doesn't he ask you to?"
"Because he's crazy," Doc Daneeka said,
"He has to be crazy to keep flying combat missions after all the close calls he's had.
Sure, I can ground Orr. But first he has to ask me to."
"That's all he has to do to be grounded?"
"That's all. Let him ask me."
"And then you can ground him?" Yossarian asked.
"No. Then I can't ground him."
"You mean there's a catch?"
"Sure there's a catch," Doc Daneeka replied,
"Catch-22. Anyone who wants to get out of combat duty isn't really crazy."

There was only one catch and that was Catch-22, which specified that a concern for one's own safety in the face of dangers that were real and immediate was the process of a rational mind. Orr was crazy and could be grounded. All he had to do was ask; and as soon as he did, he would no longer be crazy and would have to fly more missions. Orr would be crazy to fly more missions and sane if he didn't, but if he was sane he had to fly them. If he flew them he was crazy and didn't have to; but if he didn't want to he was sane and had to. Yossarian was moved very deeply by the absolute simplicity of this clause of Catch-22 and let out a respectful whistle.

"That's some catch, that Catch-22," he observed.
"It's the best there is," Doc Daneeka agreed.[22]

In evaluating reasoning, the clarification of the meaning of terms and inferences often identifies the logical fallacy. The rule in Catch-22 can be restated in the following equations:

$$\text{Catch-22 Rule} = \text{Crazy} + \text{Ask} = \text{Grounded}$$
$$\text{Ask} = \text{Sane}$$
$$\text{Sane} = \text{Not Crazy}$$

Substituting terms:

$$\text{Crazy} + \text{Not Crazy} = \text{Grounded}$$

From this equation we can see that the logical fallacy is an assumption of inconsistent terms in the first side of the equation. The contradiction is invalidly assumed in the premise of the first side of the equation.

2.3.6. *Failure to confirm hypothesis and assumed validity of alternate explanation*

A weak form of argument related to extrapolations from a small amount of data is the failure to consider other possible hypotheses. This most frequently occurs when there is a failure to look for evidence that will confirm or deny a proposed hypothesis. The uncritical acceptance of any hypothesis is not consistent with the scientific method. A similar misapplication of the scientific method occurs when one believes an alternate explanation refutes another explanation without a comparison of the merits between the two explanations. Allowing the author an indulgence in metaphysics, assume that a group of psychologists believe that they have disproved the existence of God if they have put forward a hypothesis that the idea of God is the result of projection as described in the theories of Sigmund Freud. Obviously the existence or non-existence of God is an important and difficult question, but merely finding a possible alternate explanation of the idea of God does not answer the question of God's existence. The existence of an alternate explanation does not prove much, because precisely the same argument could be given to the other side. For example, one could argue that the denial of God's existence is a projected Freudian desire to remove a person's father from his or her life. The existence of God, of course, is not an argument that can be settled by the methods of science.

I use this metaphysical example only to note the necessity to compare the merits of alternate explanations. To be consistent with the requirements of the scientific method, each hypothesis must be tested and evaluated in context. The existence of an alternate explanation may be speculation squared and sometimes speculation cubed. As we shall discuss, Stephen Hawking's hypothesis for a boundary-less universe is an alternate explanation for the hypothesis that the universe had a beginning. His hypothesis must be analyzed and evaluated in the light of issues concerning evidence for the Big Bang, temporal asymmetry and the Second Law of Thermodynamics. The mere fact that Hawking has made a speculative proposal as an alternate to a beginning for the universe does not invalidate the concept of a beginning in the Big Bang. His proposal must be examined in the light of known laws and empirically verified concepts.

2.3.7. *Confusing sequence with cause*

One must guard against confusing the sequence of events with the cause of a given event. This logical fallacy is known as *post hoc, ergo propter hoc*.[23] Because an event occurred prior to another event, one cannot conclude that the first event caused the second event. Nor can one validly assume a causal relationship between two events which occur at approximately the same time. Further, two events may be related without one being the cause of the other. A rooster's crow does not cause the rising of the sun. The fact that life exists on earth or even, if ever demonstrated, that life existed on Mars, does not prove the existence of a prebiotic soup. One passage from *King Henry IV* by Shakespeare illustrates the fallacious assumption of cause derived from sequence:

> *Glendower:*
> I cannot blame him: at my nativity
> The front of heaven was full of fiery shapes,
> Of burning cressets; and at my birth
> The frame and huge foundation of the earth
> Shaked like a coward.

> *Hotspur:*
> Why, so it would have done at the same season if your mother's cat had but kittened, though yourself had never been born.

Glendower:
I say the earth did shake when I was born.

Hotspur:
And I say the earth was not of my mind,
If you suppose as fearing you it shook.

Glendower:
The heavens were all on fire, the earth did tremble.

Hotspur:
O, then the earth shook to see the heavens on fire,
And not in fear of your nativity. (part 1, act 3, scene
1)[24]

2.3.8. Modification of question presented

Logical thinking requires a disciplined focus on the issue under consideration. One must examine evidence to be certain that it relates to the question presented and not to a slightly different question. One may use evidence to prove a conclusion, but unless the evidence relates precisely to the question at issue, the proof will only be a distraction from a valid analysis. Voltaire wrote his novel *Candide* against Leibnitz's philosophy. *Candide's* tutor, Dr. Pangloss, represented Leibnitz. Consider the following:

> Dr. Pangloss contends that this world is the best of all
> possible worlds which God could have made.
> What a ridiculous assertion! As if everything in this
> world were as good as it could be![25]

Leibnitz's philosophy was distorted by Voltaire in some respects as a theory of the "best of all possible worlds." Leibnitz's argument was that God had created the best world which it was possible for God to design. Voltaire's point against Leibnitz at first appears valid until one reviews Leibnitz's philosophy and realizes that Leibnitz did not say that everything in this world is as good as it could be, but only that this world is better than some other worlds which God could have created. Leibnitz would have argued that God could make a world without hurricanes, but then something else would have to be wrong with the world. All possible worlds considered, Leibnitz asserted that the present world is the best "possible" world.[26]

In this book the first question presented is whether, under standard probability definitions, it is mathematically possible that chance alone caused (a) the formation of the first form of living matter and (b) a universe compossible with life. The second question presented is: are any of the current self-organization scenarios for the formation of the first living matter plausible? The definition of life in this book emphasizes the sophisticated information content found in living forms of matter in the genetic code. The essential distinction between living and non-living matter is this information content which is the minimum number of instructions necessary to specify the structure under examination. If the definition of life is modified to reduce this emphasis, the examination of the questions presented is made out of context and the structure of the argument is distorted.

2.4. Limits on logic: lawyers, liars, and Gödel's Incompleteness Theorem in mathematics

Even when we employ critical thinking, validate premises and inferences, and avoid ambiguity, vagueness, and vicious circularity, human thought has severe limits; logical deduction is not perfect in its application. Some humility is required of all human beings in the context of logic's limitations. Logical systems contain imperfections.

One problem which demonstrates the imperfections in logical systems addresses the very center of the scientific method which is based upon hypothesis, investigation, and a confirming or rebutting observation. Carl G. Hempel of Princeton University wrestled with a variation of the following logical probability problem for many years: assume that a zoology professor wants to conduct a scientific investigation concerning the hypothesis that all crows are black. Using the scientific method each time she finds a black crow, she makes a notation of her discovery as a confirmation of the hypothesis.

A second hypothesis which is a logical equivalent to the hypothesis that all crows are black is the hypothesis that all non-black objects are not crows. This second hypothesis is a different ordering of words which has a meaning logically equal to the words in the first hypothesis. Because of this logically equivalent meaning, any confirmation of the second hypothesis will also confirm the first hypothesis. Each time the zoologist finds a non-black object which is not a crow she confirms the second hypothesis, which is logically a confirmation of the first hypothesis that all crows are black. For example, when she finds a red dress, she confirms the second hypothesis because the dress is not a crow.

The reasoning is validated when one considers a law firm which employs many attorneys with at least some gray hair. If a scientist wants to investigate the hypothesis that all attorneys with gray hair are married, the scientist can ask each attorney in the firm with gray hair if he or she has a spouse. A more efficient and equally valid investigation can be done by the scientist, however, by securing a list of all unmarried attorneys in the law firm and visiting each unmarried attorney and noting the presence or absence of gray hair. If none of the unmarried attorneys has any gray hair, the hypothesis that all attorneys in the law firm with gray hair are married is confirmed. Each unmarried attorney without gray hair would be a confirmation of the hypothesis that all attorneys in the law firm with gray hair are married.

Applying the same rationale to the red dress, one can assert that the discovery of a red dress is a confirmation of the second hypothesis that all non-black objects are not crows which in turn confirms the logically equivalent first hypothesis that all crows are black. Of course, there is a large disproportion between the number of crows and the number of non-black objects. Investigating all non-black objects is not an efficient method for scientific investigation. Nevertheless, the discovery of a red dress is in some sense a logical confirmation in probability of the first hypothesis. The limits of logic are apparent when one realizes that the discovery of a red dress is also in probability, *pari ratione*,[27] a confirmation of a third hypothesis that all crows are yellow. The obvious logical conundrum then becomes how the discovery of one item can act as a confirmation of the probability of two contradictory hypotheses (i.e. all crows are black versus all crows are yellow). At first review the reader may want to dismiss Hempel's problem as a curious anomaly, but upon reflection one will find the logic valid and the scientific method perhaps imperfect. Under that method the discovery of one item results in the confirmation of the probable truth of two logically inconsistent hypotheses.[28]

The limits on logic are more clearly evident in the following example: assume that Yale Law School agrees to accept Bill, a student with no money living in abject penury. Yale agrees to teach Bill the law without any tuition fees on the condition that Bill will begin to repay Yale as soon as he has won his first case. After graduating from the law school and passing the bar examination, Bill decides to teach. After several years, Yale becomes impatient and asks Bill to begin repaying his tuition. Bill responds that he was only required to begin repayment after he had won his first case and that this event had not yet occurred. Yale sues Bill in the Connecticut state court, asking the court to require Bill to begin his repayments.

In the Connecticut courtroom Bill and Yale both present sound logical arguments which validly lead to contradictory conclusions. Yale argues as follows:

(1) If Yale wins the lawsuit, then Bill must begin his tuition repayments.
(2) If Yale does not win the lawsuit, then Bill has won his first case.
(3) If Bill has won his first case, then he must begin his repayments.
Therefore, Bill must begin his repayments.

Bill learned well at Yale and argues as follows:

(1) If Yale does not win the lawsuit, then he does not have to begin his tuition repayments.
(2) If Yale wins the lawsuit, then Bill has not won his first case.
(3) If Bill has not won his first case, then he does not have to begin his repayments.

Therefore, Bill does not have to begin his repayments.

Of course, the Connecticut court could insert other factors into the argument such as a finding that the agreement implied Bill's intent to practice law and by teaching and not practicing, he violated the agreement. Or the court could find that the agreement was not binding. But these resolutions are extraneous to the valid deductive logic which results in contradictory conclusions. The facts and deductive reasoning present a logical paradox.

This logical paradox is a variation on the liar's paradox contained in the statement by Epimenides, a Cretan, who asserts, "All Cretans are always liars." If one assumes that Epimenides is telling the truth, then he is lying. But he cannot be lying because we have assumed that he is telling the truth. Similarly, if Socrates asserts, "What Plato says is a lie," and if Plato responds, "What Socrates says is true," we are faced with another logical paradox. If what Plato says is a lie, then Socrates's assertion is true. But if Socrates's assertion is true, then Plato's response is a lie and Socrate's assertion must be false.[29]

Logical difficulties are also inherent in mathematical systems. For over three hundred and fifty years Fermat's Last Theorem was thought to be one of these difficulties. Diophantus of Alexandria was a third century Greek mathematician who was fascinated by the problem of finding right triangles the length of the sides of which were in whole number ratios to one another. He wrote a treatise on this problem and others in number theory entitled, *Arithmetica*, which referred to this problem in the form of an equation splitting a sum into two squares, and asserted that the problem had no solution for cubes or hypercubes. Thus, for the equation $a^n + b^n = c^n$, no solution exists when a, b, c and n are all positive integers and n is greater than 2. An infinite number of solutions exist when $n = 2$. Pierre de Fermat, a seventeenth century lawyer and the premier mathematician of his generation, wrote in the margin of his copy of *Arithmetica*: "For this I have a marvelous proof for which this margin is too narrow."[30] No one for several centuries was able to uncover or determine a proof for Fermat's Last Theorem. Finally, in May, 1995, Dr. Andrew Wiles of Princeton University and Richard Taylor of Cambridge University published a 130-page proof of the theorem in the *Annals of*

Mathematics. This monumental task laid to rest the question whether the truth of the theorem was capable of proof or decidability. Are all mathematical propositions capable of proof or decidability given sufficient time and intelligent thought?

The answer to that question is clearly negative, no matter how much time is available. In the 1920's, German mathematician David Hilbert proposed a formalist foundation of mathematics. The purpose of Hilbert's program was to formalize all mathematics and determine proofs for the consistency of mathematics. Hilbert attempted to reduce mathematics to an axiomatic, formal system containing no contradictions and capable of demonstrating the truth or falsity by valid, logical mathematical inferences from the axioms. Hilbert tried to represent mathematical statements with the language of formal axiomatics, using the symbols of propositional and predicate calculus. In these symbols ⇒ represents "if-then," ↔ represents "if and only if," ∨ represents "or," ∧ represents "and," ~ represents "not," ∃x stands for "there exists an x such that," and ∀x stands for "for every x it is true that." Hilbert's concept of formalized mathematics permitted any proof to be expressed as a series of inferences from mathematical axioms.

In 1931, however, Austrian mathematician, Kurt Gödel, using Hilbert's expressions of formal axiomatics, demonstrated that for any consistent mathematical system there exists within the system a well formed statement that is not provable under the rules of the system. Gödel's Incompleteness Theorem is in part due to his analysis of Bertrand Russell's writings. British philosopher Bertrand Russell attempted to resolve the liar's paradox by rejecting all statements that produce vicious circularity as meaningless and neither true nor false. Kurt Gödel took Russell's monumental *Principia Mathematica* and designated a number for each one of Russell's symbols and then produced mathematical formulations of Russell's concepts. Gödel intended to prove that Russell's system was free from logical contradictions. However, Gödel discovered that to prove Russell's system consistent, he had to be able to demonstrate that any formula is or is not provable within the system. Instead of confirming Russell, Gödel developed his incompleteness theorem and demonstrated the impossibility of proving all true statements within a formal logical system.

Gödel used his designated numbering system to demonstrate that in any consistent deductive system a valid statement exists that is not provable by the rules of the system. In a mathematical system there are mathematical statements which are true but cannot be proven by the logical proofs of the mathematical system. Similarly, there are statements that are false but not provable. Gödel demonstrated that in any deductive system there is a sentence which asserts, "This sentence is not provable." Gödel was again faced with the contradictions in the liar's paradox.[31]

Gödel's theorem demonstrates that mathematics is incomplete because the system leaves unanswered the truth or falsity of certain mathematical propositions which are the logical results of valid mathematical inferences.[32] This theorem shook mathematics and all formal theories which include the arithmetic of natural numbers. If consistency could not be demonstrated within a mathematical system, at any moment a contradiction could arise and shake the system down to its foundations. There is clearly a limit on the ability of human reasoning to know that logical thought processes will lead to truth.

The limits of our reasoning powers raise the question whether scientific explanations for the origin of the laws of physics, the Big Bang, or the origin of life are issues which fall into the category of decidability represented by Fermat's Last Theorem or the indeterminate category represented by Gödel's Incompleteness Theorem. For over three hundred and fifty years mathematicians were unable to prove or disprove Fermat's Last Theorem. Yet in 1993 the proof was finally accomplished. But Gödel's Incompleteness Theorem raises an intractable or indeterminate problem which will always be beyond the powers of human reason. The missing ingredient in all scientific origin of life scenarios is an explanation of a method for generating sufficient information content into inert matter to qualify that matter as living. In our examination of the concept that chance alone was responsible for life, we will also consider some other self-organization scenarios which also fail to supply this missing ingredient. To make progress on an adequate theory, science needs to remove all failed scenarios and concentrate on the method by which information content could be generated into inert matter. Only by removing the failed models will we be able to determine whether the propositions are decidable or indeterminate. Similarly, "pre" Planck time (10^{-43} of the first second) activity may always remain beyond our understanding, because the known laws of physics break down at Planck time. Science must attempt to find answers to these questions, but we may conclude that the questions are intractable. Time will tell whether we will finally discover the answer as Wiles did for Fermat's Last Theorem or conclude that these questions are not decidable and belong to the category of propositions found in Gödel's Incompleteness Theorem.

2.5. Uncertainty in quantum mechanics

Four years before Gödel demonstrated the need for humility in evaluating deductive reasoning, physicist Werner Heisenberg discovered a principle which demonstrated the limitations of knowledge in quantum physics. Heisenberg's principle states that it is impossible to measure simultaneously the precise position and momentum of an elementary particle. We will always be uncertain about one or the other. The equation for the principle is $\Delta x \Delta p_x \geq h/4\pi$, where Δx is the uncertainty in the position of the particle, Δp_x is the uncertainty in the particle's momentum (mass times velocity), and h is the Planck constant (a fundamental value equal to the ratio of the energy of a quantum to its frequency).[33] The equation demonstrates that there is an unavoidable uncertainty in the product of position and momentum. To locate the precise position of an electron, an observer must be able to strike the electron with light photons, but this act randomly changes the electron's position. What one observes depends to some extent on how one observes. The observer cannot be removed from the subject of the observation.

Despite the uncertainty that exists at the quantum level and despite the healthy humility required by Gödel's theorem, on a larger physical scale and in a pragmatic manner, we find that the principles of mathematics and logic work extremely well in the physical world. On this larger scale Einstein's famous assertion rings true: *"Der liebe Gott würfelt nicht mit der Welt."*[34] We all trust these principles every time we drive a car or fly in an airplane. Mathematics work remarkably well in describing the physical world and in application to the many electrical, acoustical, optical, and mechanical products we use every day. Although human beings are imperfect in their reasoning and in their observation, logical thought, observation, and mathematical analysis are the best instruments we have available in reviewing the evidence for and against accident in the formation of the universe and in the origin of life.

PART III

CASE AGAINST ACCIDENT FROM MATHEMATICAL PROBABILITIES IN MOLECULAR BIOLOGY

In this section we will begin with an accepted definition of life which emphasizes the information content required in a living system. We will then review the theoretical model for the emergence of life by chance processes and examine the reliability of the Miller and Urey line of experiments. We will calculate the very limited time available for the formation of life on earth. Since proponents of the origin of life by accident or chance processes rarely make the mathematical calculations of the probabilities which lie at the foundation of their hypothesis, we will discuss the process of calculating the probability of an accidental or chance event and review the calculations of many noteworthy scientists. Because the odds are so overwhelmingly against the formation of life from accidental or chance processes, in Part IV we will discuss some theories which emphasize self-organization scenarios in systems far from equilibrium. The examination of these proposals will include the problem of finding a plausible method of generating sufficient information content, such as that represented by the genetic code, into inert matter. Finally, we will discuss findings concerning the Allan Hills 84001 meteorite and their effect, if any, on the questions presented in this book.

3.1. Definition of life

While practicing in my law firm's Washington, D.C. office in 1978, I taught on the weekends as part of the faculty of the University of Virginia School of Law. My course was an advanced course in secured financing for third year law students. The Uniform Commercial Code was an important part of my course. To encourage uniformity in interpretation and to enhance their understanding of the subject matter, I urged the students to study carefully the terms defined in the Code. A criticism of the Code, encountered frequently, was that the Code did not define certain fundamental concepts such as "commercially reasonable." The reason in part was the difficulty in articulating these concepts except by the process of describing examples of conduct which conform to commercial reasonableness. An attempt to define life encounters analogous problems. Defining life is similar to defining time; we all experience it, but have difficulty in describing its precise characteristics. Living systems such as human beings are made up of cells, each containing all the characteristics of life. These cells in turn are made up of atoms with about as many atoms in a cell as there are cells in a human being. Individual atoms are not alive, and the dividing line between living and dead matter appears to be somewhere between a cell and an atom.[35] George Gaylord Simpson, the highly regarded professor of paleontology at Harvard University, gave the following definition of a living system:

> A fully living system must be capable of energy conversion in such a way as to accumulate negentropy, that is, it must produce a less probable, less random organization of matter and must cause the increase of available energy in the local system rather than the decrease demanded in closed systems by the second law of thermodynamics. It must also be capable of storing and replicating information, and the replicated information must eventually enter into the development of a new individual system like that from which it came. The living system must further be enclosed in such a way as to prevent dispersal of the interacting molecular structures and to permit negentropy accumulation. At the same time selective transfer of materials and energy in both directions between organism and environment must be possible. Systems evolving toward life must become cellular individuals bounded by membranes.[36]

Living matter processes energy, stores information and replicates. To be alive a system must achieve a certain level of complexity to perform these functions. A central distinction between living and non-living matter is the existence of a genome or composite of genetic messages which carry sufficient information content to replicate and maintain the organism. To be considered alive an organism must have enough information content to control its genetic and biochemical processes. The requirement of this information content is indispensable to an adequate definition of life for reasons which will be discussed in the section of this book examining propositions relating to the development of life in open systems far from equilibrium.

In attempting to determine the level of complexity or information content required to consider a structure alive, we may look at a bacteriophage ("phage") which is a virus that is a parasite within a bacterium. Each type of phage is unique to each type of bacterium. Bacteria are single living cells which do not combine to form a more complex living system. Phages are not considered to be alive, because they cannot exist or replicate themselves without assistance from their host bacteria. The dividing line for complexity and information content sufficient to call a structure life appears to be somewhere between a phage and a bacterium.[37]

Recent discoveries in molecular biology portray the enormous complexity in the smallest living cell. A single-celled bacterium contains millions of atoms and an enormous number of informational instructions. Michael Denton describes this complexity:

> Molecular biology has shown that even the simplest of all living systems on earth today, bacterial cells, are exceedingly complex objects. Although the tiniest bacterial cells are incredibly small, weighing less than 10^{-12} gms, each is in effect a veritable micro-minaturized factory containing thousands of exquisitely designed pieces of intricate molecular machinery, made up altogether of one hundred thousand million atoms, far more complicated than any machine built by

man and absolutely without parallel in the non-living
world . . . The recently revealed world of molecular
machinery, of coding systems, of informational molecules, of catalytic devices and feedback control, is in
its design and complexity quite unique to living systems and without parallel in the non-living world.[38]

3.2. DNA, RNA, protein synthesis and the genetic code

Because the central distinction between living and non-living
matter is the presence of sufficient information content to control
the genetic and biochemical processes of the living matter, a brief
review of the system of information transfer through the genetic
code is essential. All living matter contains DNA (deoxyribonucleic acid) as its genetic material. DNA stores and transfers the information required for life processes. One function is the transfer
of genetic information through the replication process. Another
function of DNA is the transfer of information needed to form
the specific enzymes necessary for the functions of the organism's cells.

A DNA molecule is comprised of thousands of long chains of
nucleotides (polynucleotides) each consisting of three parts. One
part is the pentose or five carbon sugar known as deoxyribose. A
second part is a phosphate group, and the third part is a nitrogen
base of either adenine (A), guanine (G), cytosine (C) or thymine
(T). Alternating sugar and phosphate molecules connect each nucleotide chain in a ladder type configuration coiled around a central axis in a twisted double spiral or helix. The two chains run in
opposite directions with 10 nucleotides per turn of the helix. The
rungs of the bases are pairs of either adenine and thymine (A-T)
or cytosine with guanine (C-G). A relatively weak hydrogen bond
connects these bases. The process of replicating this DNA molecule begins with the breaking of the hydrogen bond and the
splitting in two of the spiral ladder. Free nucleotides present in
the cell bond with matching nucleotides in the half ladder structure with bases matching. (A with T and C with G). A new up-
right portion is added as the once split half ladder becomes a
replication of the initial DNA molecule.

DNA also contains the information content or instructions for the formation of enzymes, specialized proteins that serve as catalysts in a cell's chemical reactions. These instructions are sent through RNA (ribonucleic acid) which is also comprised of nucleotides with a five carbon sugar known as ribose. RNA nucleotides are arranged in a single strand with nitrogen bases of uracil (U), adenine (A), cytosine (C), and guanine (G). The four bases of RNA and the four bases of DNA comprise the symbols or alphabet of the genetic code. A code, of course, is a system of symbols like the Roman alphabet of the English language or the dots and dashes of the Morse Code which correspond to that alphabet. Proteins are built from twenty different amino acids which means that each amino acid must receive a distinct message so that the instructions from DNA must be capable of forming at least twenty combinations. The four bases of the DNA nucleotide sequence combine in groups of three to give 64 possible triplet combinations. These triplet combinations are called codons. Each codon carries a precise set of instructions.

Protein or enzyme synthesis results from the combination of amino acids pursuant to the instructions from DNA. This synthesis begins with the initiation of the transcription process, when RNA polymerase, the transcription enzyme, binds to DNA at the promoter region (which promotes the transfer of instructions in the transcription). RNA synthesis begins with the unraveling of the DNA spiral ladder. One of the strands of the DNA acts as a coding strand and directs the synthesis of the RNA. The RNA polymerase reads the DNA coding sequence and moves along the DNA, with the new synthesized strand of the RNA transcript molecule elongating until the RNA polymerase reaches the end of the coding region. At this point of termination the RNA polymerase and the newly synthesized RNA, known as the primary RNA transcript, are released from the DNA.

Before leaving the nucleus of the cell, the primary RNA transcript is processed into a mature messenger RNA (mRNA). The new messenger RNA moves through a pore in the nuclear membrane, enters the cytoplasm and locates in a ribosome, a small spherical body within the cell and the site of protein synthesis. mRNA communicates the necessary instructions from DNA in the nucleus to the ribosome. These instructions are the genetic code carried in sequences of codons. Thus, the genetic code is written in mRNA with each codon designating an amino acid. The amino acid symbols (given in parentheses) for the 20 amino acids used in living matter are: Alanine (Ala), Arginine (Arg), Asparagine (Asn), Aspartic acid (Asp), Cysteine (Cys), Glutamic acid (Glu), Glutamine (Gln), Glycine (Gly), Histidine (His), Isoleucine (Ile), Leucine (Leu), Lysine (Lys), Methionine (Met), Phenylalanine (Phe), Proline (Pro), Serine (Ser), Threonine (Thr), Tryptophan (Trp), Tyrosine (Tyr), and Valine (Val).

The following table gives the codons for the amino acids. Note that the AUG codon gives instructions to start and the codons UAA, UAG, and UGA give instructions to stop.[39]

First Letter	Second Letter				Third Letter
	U	C	A	G	
U	phenylalanine	serine	tyrosine	cysteine	U
	phenylalanine	serine	tyrosine	cysteine	C
	leucine	serine	stop	stop	A
	leucine	serine	stop	tryptophan	G
C	leucine	proline	histidine	arginine	U
	leucine	proline	histidine	arginine	C
	leucine	proline	glutamine	arginine	A
	leucine	proline	glutamine	arginine	G
A	isoleucine	threonine	asparagine	serine	U
	isoleucine	threonine	asparagine	serine	C
	isoleucine	threonine	lysine	arginine	A
	(start)methionine	threonine	lysine	arginine	G
G	valine	alanine	aspartate	glycine	U
	valine	alanine	aspartate	glycine	C
	valine	alanine	glutamate	glycine	A
	valine	alanine	glutamate	glycine	G

Pursuant to the genetic code, transfer RNA (tRNA) binds with specific amino acids in the cytoplasm and brings them to two binding sites in the ribosome. At these sites, the mRNA codons base pair with the tRNA anticodons which are also comprised of groups of three nucleotides. For example, if tRNA contains an anticodon with the sequence UGC, it will fit the mRNA codon ACG. Uracil (U) always bonds with adenine (A), and cytosine (C) always bonds with guanine (G). The codons dictate the amino acid sequence as they grow into a polypeptide chain. Thus, a new protein is synthesized and ready to perform its function in the living organism. This process of transforming the genetic code instructions from mRNA into a protein is known as translation.

One may think of the genetic code process as similar to the production of a novel. The DNA instructions act like sentences. These sentences are copied into a message in mRNA which moves to a printing facility called a ribosome. The message is written in three letter words which are codons, and the printing facility produces a book called a protein. The analogy is useful but imperfect, because the protein is ready to act and perform functions, and a book is passive and only ready to be read. The key concept is that an indispensable characteristic of living matter is a complex content of information sufficient to replicate and maintain the organism.

The information contained in the genetic code, like all information or messages, is not made of matter. Materialism does not explain the meaning in the code. The meaning is not a property of the arrangement of the symbols or alphabet of the code. The message or meaning in the genetic code is non-material and cannot be reduced to a physical or chemical property. Hubert Yockey, an erudite physicist who studied under J. Robert Oppenheimer at Berkeley and then worked with him on the Manhattan Project, uses the analogy among letters of the Roman alphabet and their meaning in the English, French and German languages to demonstrate the non-material nature of the messages and information in the genetic code:

> . . . the meaning, if any, of words, that is, a sequence of letters, is arbitrary. It is determined by the natural language and is not a property of the letters or their arrangement. For example, the English word "hell" means bright in German, "fern" means far, "gift" means poison, "bald" means soon, "boot" means boat, "singe" means sing. In French "pain" means bread, "ballot" means a bundle, "coin" means a corner or a wedge, "chair" means flesh, "cent" means hundred,

"son" means his, "tire" means a pull, "ton" means your. This confusion of meaning goes as far as sentences. For example, "O singe fort!" has no meaning as a sentence in English, although each is an English word, yet in German it means "O sing on!" and in French it means "O strong monkey". Like all messages, the life message is non-material but has an information content measurable in bits and bytes and plays the role, ascribed by vitalists, of an unmeasurable, metaphysical vital force without being *ad hoc*, romantic, spooky, contrary to the laws of physics or supernatural. Of course, like all messages, the genetic message, although non-material, must be recorded in matter or energy.[40]

3.3. Theory of emergence of life from accidental or chance processes

A theoretical model for the emergence of life was proposed by Soviet scientist Alexander Oparin in 1924[41] and British chemist J. B. S. Haldane in 1928.[42] Oparin and Haldane based their theory on the early earth's atmosphere consisting of methane, ammonia, carbon monoxide, carbon dioxide, hydrogen and water vapor with ultraviolet light acting upon this atmosphere. This type of atmosphere is known as a "reducing" atmosphere without oxygen. According to their theory, energy sources acting on the early earth's atmosphere resulted in the formation of organic compounds in the atmosphere which were washed down by rain and accumulated in the primitive oceans until they reached the consistency of a hot dilute soup. According to this model, life appeared from the chemical reactions and transformations that took place in this prebiotic soup.

A brief review of an example of this model may be useful. The hypothesis is basically as follows: Volcanic eruptions formed the early earth's atmosphere which contained the following: carbon dioxide, ammonia, methane, hydrogen, water vapor and no oxygen. These compounds were washed by rain into the hot oceans where they were subject to energy charges from ultraviolet radiation, lightning, shock waves, and heat. These energy charges caused the compounds to form chemical bonds with dissolved minerals, producing amino acids and sugars. These amino acids combined to form peptides of two or more amino acids linked by peptide bonds (formed by reactions between adjacent carboxyl and amino groups with the elimination of water). The peptides combined to form polypeptides (a peptide with ten or more amino acids) and proteins. These organic molecules became more complex and formed clusters of molecules capable of heterotrophism which is a type of nutrition in which energy results from the intake of organic substances (as opposed to autotrophism which is a type of nutrition in which organisms synthesize organic materials from inorganic sources). These heterotrophs increased in complexity, and nucleic acids were formed which gave them the ability to reproduce. At this point the heterotrophs were alive. They had a form of anaerobic respiration which converted pyruvate (the end product of glycolysis, the breaking down of glucose) into ethanol and carbon dioxide. Certain of the heterotrophs evolved a method of using carbon dioxide to synthesize organic materials from inorganic sources and became autotrophs. These autotrophs produced oxygen. The heterotrophs and autotrophs then evolved methods to use aerobic respiration to employ oxygen to secure energy in the nutrition process.

Whether life emerged in a gradual manner is not the principal issue raised in this book. Let us assume for a moment that life emerged from a gradual scenario; the first question presented is whether the origin was guided, or accidental and by chance. In his book, *Chance and Necessity* (1971), Nobel laureate Jacques Monod asserted that blind chance was the reason for all forms of life:

> Chance alone is the source of every innovation, of all creation in the biosphere. Pure chance, absolutely free but blind, is at the very root of the stupendous edifice of evolution. The central concept of biology . . . is today the sole conceivable hypothesis, the only one compatible with observed and tested fact. All forms of life are the product of chance . . . [43]

As we shall demonstrate, the mathematical probability calculations in this book run directly counter to any support for Monod's assertion.

In the 1950s Stanley Miller and Harold Urey simulated portions of the Oparin and Haldane description of the early earth's atmosphere in a laboratory apparatus at the University of Chicago. The apparatus was sterilized and airtight, containing methane, ammonia, and hydrogen gases which circulated past a high-energy electrical spark. The apparatus was connected to a container of boiling water which supplied heat and water vapor. When the water vapor circulated through the apparatus, it condensed. Accordingly, Miller created what he believed to be some of the conditions of the early earth's atmosphere: gases, heat, rain (condensed water vapor) and flashes of lightning (the energy spark), as they were believed to have occurred in the early atmosphere. After the gases had circulated for a week, Miller saw a small mass of black tar in one part of the apparatus and a condensed red liquid. The analysis of the liquid disclosed some amino acids which are the building blocks of protein.

Other experiments based on the Miller-Urey assumptions modified the simulated atmosphere conditions and used ultraviolet radiation to produce nineteen of the twenty biological amino acids and five nucleic acid bases of DNA and RNA used in the genetic process. These experiments did not prove that amino acids formed in this way under the early conditions of the earth, but the widely held scientific view was that a similar random process somehow produced a simple form of life *de novo* out of inorganic substances. Random abiogenesis became the accepted theory in college textbooks despite the absence of evidence supporting this view.

As we shall discuss, discoveries in molecular biology and in the geological records raise profound doubts about this view and the relevance of the Miller-Urey and other monomer[44] experiments. The information filled molecules of life are much more complex and structured than previously thought, and calculations of the mathematical probabilities of unguided, chance processes forming life call the theory of accidental abiogenesis into question.

3.4. Facticious flaws in the Miller and Urey line of experiments

Miller and Urey's experiment only works as long as oxygen is absent and certain critical ratios of hydrogen and carbon dioxide are maintained. Only in these conditions are amino acids produced. Scientists are now learning that the atmosphere of the early earth probably was not of the strongly reducing nature required by the Miller-Urey apparatus. Oxygen was likely present in the early earth's atmosphere. Of course, when one speculates about the origin of life, he or she has no way of knowing and scientifically verifying the actual conditions of the early earth. We are left to analyze what is plausible given our present data and our understanding of the laws of the physical world.

3.4.1. *Less reducing atmosphere of early earth*

Miller and Urey's experiment and other subsequent similar monomer experiments assumed a strongly reducing atmosphere for the early earth without oxygen. Scientists today are of the opinion that the earth's primitive atmosphere was not so strongly reducing and probably contained significant amounts of oxygen.[45] The presence of even a small amount of oxygen, assiduously avoided in the laboratories of these experiments, would prevent the formation of amino acids and nucleotides, because atoms and molecules would bond with the oxygen atoms rather than hydrogen atoms. Even if amino acids could be formed, oxygen would cause them to decompose quickly and terminate any further random processes which could eventually produce life. If the early earth's atmosphere had oxidizing conditions, abiogenesis would have been impossible.[46] R. T. Brinckmann calculated the amount of oxygen generated from photodissociation and consumed in the oxidation of rock. His analysis indicated that 25% or more of the present level of oxygen existed over 99% of the time since the formation of sedimentary rocks. He concluded that chemical evolution could not have proceeded in such an atmosphere.[47] J. H. Carver, a scientist at the Research School of Physical Sciences at the Australian National University, calculated the quantity of oxygen produced by photodissociation in the early earth's atmosphere and wrote that the concentration of free oxygen could have reached 10% of the present level which would also prevent the formation of amino acids.[48] The most recent data indicates a trend in assuming more oxygen than contemplated by the Miller and Urey experiment:

> The only trend in the recent literature is the suggestion of far more oxygen in the early atmosphere than anyone imagined. A significant part of this trend is due to measurements which suggest that stars resembling the sun at a few million years of age emit up to 10^4 times more UV light than the present sun. This increase in UV could increase the O_2 surface mixing ratio by a factor of 10^4 to 10^6 over the standard value of 10^{-15}, thus affecting all the oxygen level estimates. Support for large estimates of O_2 is found in data from Apollo 16—data which suggest that a large amount of free oxygen does result from upper atmosphere photodissociation of water vapor.[49]

Even if oxygen was not present in the early earth's atmosphere, the absence of oxygen would present obstacles to the formation of life. Oxygen is required for the ozone layer which protects the surface of the earth from deadly ultraviolet radiation. Without oxygen this radiation would break down organic compounds as soon as they formed. This lethal ultraviolet flux is part of the Catch-22 against abiogenesis. As Michael Denton notes:

> What we have then is a sort of "Catch 22" situation. If we have oxygen we have no organic compounds, but if we don't have oxygen we have none either. There is another twist to the problem of the ultraviolet flux. Nucleic acid molecules, which form the genetic materials of all modern organisms, happen to be strong absorbers of ultraviolet light and are consequently particularly sensitive to ultraviolet-induced radiation damage and mutation. As Sagan points out, typical contemporary organisms subjected to the same in-

tense ultraviolet flux which would have reached the Earth's surface in an oxygen-free atmosphere acquire a mean lethal dose of radiation in 0.3 seconds . . . The level of ultraviolet radiation penetrating a primeval oxygen-free atmosphere would quite likely have been lethal to any proto-organism possessing a genetic apparatus remotely resembling that of modern organisms.[50]

A methane rich reducing atmosphere is essential to the Oparin-Haldane hypothesis and the Miller and Urey experiment. Miller based his experiment on the cosmic abundance of hydrogen and the ingredients in the solar nebulae which he believed produced the early earth's atmosphere. The current geological consensus, however, maintains the view that the interior earth, rather than the solar nebulae, produced the primitive atmosphere and that methane and ammonia were not present. Today geologists understand that chemical reactions from sunlight would have destroyed methane and ammonia within a few thousands years. The sun's ultraviolet radiation would have converted the methane to hydrocarbons with higher molecular weight and formed an oil slick up to a depth of ten meters.[51] Ammonia is destroyed by ultraviolet radiation, dissociating into nitrogen gas and hydrogen. This presents a stumbling block for anyone building his or her theory of the origin of life on the Oparin-Haldane foundation. As Miller himself admitted: "If it is assumed that amino acids more complex than glycine were required for the origin of life, then these results indicate a need for CH_4 (methane) in the atmosphere."[52]

Yet many scientists hang on to the myth of a strongly reducing environment. Manfred Schidlowsky stated: "The very fact that life sprang up on earth constitutes conclusive proof of a primary reducing environment since the latter is a necessary prerequisite for chemical evolution and spontaneous origin of life."[53] This is a good example of the circular reasoning discussed above in which evidence is ignored in order to maintain a myth, and the conclusion is set forth in the premise.

Robert Shapiro borrows the description of Gunnar Sillen and describes the Oparin-Haldane hypothesis as "the myth of the prebiotic soup."[54] He agrees with Carl Woese's criticism of the current dogma taught in a majority of biology and biochemistry textbooks:

> The Oparin thesis has long ceased to be a productive paradigm: It no longer generates novel approaches to the problem; more often than not it requires modification to account for new facts; and its overall effect now is to stultify and generate disinterest in the problem of life's origin. These symptoms suggest a paradigm whose course is run, one that is no longer a valid model of the true state of affairs.[55]

3.4.2. Inefficacy of random distribution of left and right handed molecules as building blocks for life

Miller and Urey's experiment produced a random distribution of left and right handed molecules. Amino acids are in one of two forms: L-amino acids (left-handed molecules) or D-amino acids (right-handed molecules), each a mirror image of the other. Only left-handed amino acids (L-amino acids) are contained in biologically functional proteins. None of the acids produced in the experiment combined with each other in any way. For protein functions amino acids must combine in a sophisticated sequence. This sequence is not easy to obtain by random processes, because L-amino acids and D-amino acids bond without distinction, and D-amino acids and L-amino acids are equally present in the physical world. Forming a sequence of only L-amino acids is necessary for the formation of a protein with enzymatic functions necessary for life.

3.4.3. Dilution processes in the prebiotic soup and the prevention of formation of polypeptides

In his book *Origins: A Skeptic's Guide to the Creation of Life on Earth*, Robert Shapiro reviews the evidence for the concentrations of amino acids in the early ocean. After examining the processes as proposed, including the degradation of amino acids by ultraviolet radiation as they circulated to a depth of tens of meters near the ocean surface, he concluded that the hypothesized prebiotic soup never existed. Processes of dilution would have prevailed in the hypothesized prebiotic soup, greatly diluting the amount of precursor chemicals and preventing the formation of polypeptides. Solar ultraviolet light, thermal conditions, lightning, shock waves, and the hydrolysis of hydrogen cyanide and nitriles would have destroyed many of the organic compounds in the ocean. Even if polypeptides[56] had formed in the primordial soup, hydrolysis[57] would have broken them up and destroyed most amino acids.[58] Michael J. Behe, a professor of biochemistry at Lehigh University, comments on the effect of hydrolysis in preventing the formation of polypeptides:

> . . . joining many amino acids together to form a protein with a useful biological activity is a much more difficult chemical problem than forming amino acids in the first place. The major problem in hooking amino acids together is that, chemically, it involves the removal of a molecule of water for each amino acid joined to the growing protein chain. Conversely, the presence of water strongly inhibits amino acids from forming proteins. Because water is so abundant on the earth, and because amino acids dissolve readily in water, origin-of-life researchers have been forced to propose unusual scenarios to get around the water problem.[59]

All of the experiments producing amino acids reported a tarry substance as the major product of the experiment. If the primordial soup existed, this tarry substance must have been prevalent before the emergence of the first life form and should be found at least as a non-biological kerogen in the geological records. Yet there is no trace of any such substance.[60]

Urey and Miller assumed that methane was plentiful in the early earth's conditions. If this is true, the sun's ultraviolet light would have caused hydrocarbons to form and adsorb in the clay at the bottom of the ocean. The deposits from Precambrian periods should then contain significant hydrocarbons or remains of carbons, as well as some nitrogen containing compounds. None of these are present in these deposits. Their absence has been emphasized by scientists attempting to ascertain the plausible conditions on the primitive earth:

> If there ever was a primitive soup, then we would expect to find at least somewhere on this planet either massive sediments containing enormous amounts of the various nitrogenous organic compounds, amino acids, purines, pyrimidines, and the like, or alternatively in much-metamorphosed sediments we should find vast amounts of nitrogenous cokes (graphite-like nitrogen-containing materials). In fact, no such materials have been found anywhere on earth.[61]

The existence of a prebiotic soup is an essential prerequisite for the traditional theory of the emergence of life by accidental processes. Michael Denton is impressed with the absence of any evidence for such a soup in the earliest geological records:

> The existence of a prebiotic soup is crucial to the whole scheme. Without an abiotic accumulation of the building blocks of the cell no life could ever evolve. If the traditional story is true, therefore, there must have existed for many millions of years a rich mixture of organic compounds in the ancient oceans and some of this material would very likely have been trapped in the sedimentary rocks lain down in the seas of those remote times. Yet rocks of great antiquity have been examined over the past two decades and in none of them has any trace of abiotically produced organic compounds been found. Most notable of these rocks are the "dawn rocks" of Western Greenland, the earliest dated rocks on Earth, considered to be approaching 3,900 million years old. So ancient are these rocks that they must have been lain down not long after the formation of the oceans themselves . . . Sediments from many other parts of the world dated variously between 3,900 million years old and 3,500 million years old also show no

> sign of any abiotically formed organic compounds . . .
> Considering the way the prebiotic soup is referred to
> in so many discussions of the origin of life as an al-
> ready established reality, it comes as something of a
> shock to realize that there is absolutely no positive ev-
> idence for its existence.[62]

A more plausible interpretation of the evidence is that chemical reactions on the primitive earth would have rendered any
soup too dilute and made compounds unsuitable for the formation of life. The primordial soup, as contemplated by Haldane
and Oparin, probably never existed. Yet the prebiotic soup is an
absolutely required presupposition to the Miller and Urey materialist paradigm of the formation of the first life form. The Isua
sedimentary rocks in Western Greenland described by Michael
Denton are over 3.8 billion years old. Writing in *The Journal of The-*
oretical Biology, Hubert Yockey asserts that the "absence of evidence" is the "evidence of absence" for the prebiotic soup. He
analyzes the significance of the depletion of ^{13}C and the enhancement of ^{12}C in the kerogen found in the Isua rocks:

> The vastly more abundant result of all "prebiotic" ex-
> periments is an insoluble tarry mixture. After the ori-
> gin of life, this tarry mixture would have precipitated
> out of the primeval ocean and have been found in the
> kerogen of sedimentary rocks. Since it would have
> carried the ^{13}C rejected by enzymatic action, no en-
> hancement of ^{12}C would have occurred. The signifi-
> cance of the isotopic enhancement of ^{12}C in the very
> old kerogen in the Isua rocks in Greenland is that
> there never was a primeval soup and that, neverthe-
> less, living matter must have existed abundantly on
> earth before 3.8 billion years ago.[63]

In his brilliant book, *Information Theory and Molecular Biology*,
published by Cambridge University Press, Yockey also criticizes
proponents of abiogenesis for allowing their metaphysical
assumptions to override the results of experiments and mathematical analysis:

Although the Oparin-Haldane paradigm is now just a
relic of the cosmology of the time when it was invent-
ed, it certainly deserved extensive research and much
has been learned in investigating it. The same can be
said for many other failed paradigms. Nevertheless,
like the luminiferous ether, one has to conclude that
there is no evidence that a "hot dilute soup" ever existed. In spite of this fact adherents of this paradigm
think it *ought* to have existed for philosophical or ideo-
logical reasons . . . It is universally the case that textbooks written for college undergraduates present the
primeval soup paradigm as an established fact. . . . I
have emphasized that in science one must follow the
results of experiments and mathematics and not one's
faith, religion, philosophy or ideology. The primeval
soup is unobservable since, by the paradigm, it was destroyed by the organisms from which it presumably
emerged. It is most unsatisfactory in science to explain
what is observable by what cannot be observed. Since
creative skepticism and not faith is the cardinal virtue
in science one would expect that proponents of the
primeval soup paradigm would be actively searching
for direct geological evidence of such a condition of
the early ocean. The power of ideology to interpose a
fact-proof screen is so great that this has not been done
(perhaps for fear that its failure may be exposed).[64]

3.4.4. Factor of facticious manipulation of researcher

Miller and Urey did not examine random conditions. The definition of facticious used in this paragraph means a contrived manipulation by a human being. Something occurs facticiously when it is forced into being by a human agency. Miller and Urey's and several subsequent experiments were facticious in the sense that the conditions were meticulously manipulated by the researchers within the glass tubes. If researchers using their full level of scientific and technical skills are not able to form living organisms from amino acids, one must ask how life formed before this intelligence existed to manipulate the environment.[65]

There are limits to what random natural processes alone have achieved in a laboratory environment, compared to the achievements accomplished through the interference of a researcher. As Brooks and Shaw have noted: "These experiments . . . claim abiotic synthesis for what has in fact been produced and designed by highly intelligent and very much biotic man." [66] Only biotic processes direct energy flow to the work of forming life. To paraphrase Louis Pasteur, in experience only life produces life.

Despite the presentation of the Miller and Urey paradigm as tantamount to fact in most college and secondary school textbooks, the results of these experiments have not produced a plausible theory which is acceptable among most prominent origin of life scientists. The Miller and Urey line of experiments does not "work." Later in the book we will show probability calculations which demonstrate that the theory underlying these experiments is not even mathematically possible. As Hubert Yockey in his succinct style simply concludes:

> In so far as chance plays a central role, the probability that even a very short protein, not withstanding a genome, could emerge from a primeval soup, if it ever existed, even with the help of a *deus ex machina* for 10^9 years is so small that the faith of Job is required to believe it.[67]

3.5. Limited time available for formation of life from accidental or chance processes

In his article published in *Scientific American*, Nobel prize winning biologist George Wald presented his argument for chance processes and abiogenesis. "Time itself performs the miracles", he argued. "Time is in fact the hero of the plot."[68] Wald's view was widely adopted, but he wrote in 1954 without the benefit of recent discoveries in physics and the fossil records which indicate a very short period of time available for life to form inevitably from inorganic chemical processes. For the proponents of origin of life by chance, recent discoveries disclose that time is not the hero of their plot but the villain.

Haldane, Oparin and Wald wrote their papers at a time when the universe was believed to have no beginning or end and to be infinite in size. In an eternal, infinite universe, anything can happen. Data supporting the Big Bang theory from the Cosmic Background Explorer satellite and new discoveries in the geological records change the perspective of the time available for the emergence of life. The time available on earth is extremely limited. The earth began to form about 4.6 billion years ago. Radioactive decay, the greenhouse effect in the atmosphere, the production of thermal energy from the effects of gravity conversion, and crashing meteors made the surface of the earth sufficiently hot to make compounds of biological interest unstable for approximately 1.62 billion years.[69] In other words, prior to 3.98 billion years ago, the earth was too torrid for the emergence of life. The fossil records, however, indicate that life formed on earth at least 3.85 billion years ago or over a period of less than 130 million years.

Itasq is a word in the Greenlandic language which means "ancient thing."[70] The Itasq Gneiss Complex, a geological area in west Greenland, contains the planet's oldest known fossil records. These records indicate that life began on earth almost immediately upon sufficient cooling (approximately 100°C).[71] A banded iron formation from the Isua supracrustal belt of western Greenland and a similar formation from the nearby Akilia Island contain what may be the oldest evidence of life on earth dating back to at least 3.85 billion years ago.[72] Recently a six member team, headed by S. J. Mojzsis of the Scripps Institution of Oceanography, reported ion-microprobe measurements of the carbon-isotope composition of carbonaceous inclusions within grains of apatite in sediment from Akilia. They concluded that the carbon was isotopically light, indicative of biological activity which could not be explained by any abiotic process:

> We therefore conclude that metamorphic effect are not responsible for the association of isotopically light carbonaceous inclusions in metasedimentary apatite. Together with the intergrowth of carbonaceous matter with apatite in BIF (banded iron formation) from Akilia Island, we conclude that the isotopic results re-

ported here give strong evidence for life on Earth by
3,850 Myr.[73] (parentheses added)

*In other words, only a maximum of 130 million years were available
for random processes to produce life.* Calculations of mathematical
probabilities unequivocally demonstrate that it is mathematically
impossible for unguided, random events to produce life in this
short period of time.

A. G. Cairns-Smith, in addition to concluding that the prebiotic soup never existed, points out that even a period of 200 to
300 million years is far too short for any living organism to form
from random processes.[74] On this point Harold Morowitz agrees
with Cairns-Smith. In commenting on the low probability of random events forming life in such a short window of time, Harold
Morowitz wrote:

> I think it is conservative to say that continuous life on
> Earth formed 3.8 ± 0.2 Ga ago. This is not a precise es-
> timate, but it places the event in the late Hadean or
> early Archean period, suggesting that as soon as the
> Earth cooled down sufficiently, life formed rapidly on
> a geological time scale. A less conservative estimate
> would be $3.9. \pm$ Ga ago—a very different view from
> the classical perspective involving random chemicals
> reacting for eons and finally lucking out, resulting in a
> living cell coming together. The thrust of narrowing
> the window in time is to shift the emphasis from low
> probability random events to the deterministic pro-
> duction of living entities.[75]

3.6. Calculating mathematical probabilities
of accidental or chance events

Many proponents of the origin of life by chance do not bother to
perform the mathematical calculations which render their conclusions highly improbable. To determine probabilities for the accidental formation of life, one cannot rely on intuition. Intuition
in probability theory is not a very accurate guide; one can be led

astray if all outcomes are not considered. Much of the material in this book centers on the sequences of nucleotides and amino acids which carry the information of the genetic code. These information bearing sequences are part of living matter but not part of non-living matter. In assessing probabilities for the formation of such sequences, all outcomes must be considered.

To appreciate the need for reviewing all possible outcomes, and not relying on intuition, assume that there are three boxes each containing colored wooden balls. The first box contains two red balls; the second box contains two green balls; and the third box contains a red and a green ball. If you blindfold friend X and ask X to select one of the boxes, the probability that he will select the box with two green balls is .3333. If you ask X to select one of the balls out of a box and this ball is green, your intuition may be that the probability that the remaining ball is red is .5. This is not the correct probability. An examination of possible outcomes will demonstrate the correct probability.

The green ball came from either the second box with two green balls or the third box with a red and a green ball. If X had selected the second box, the probability of X selecting a green ball would be 1.0. If X had selected the third box, the probability that the remaining ball is also green is .5. When X selected the green ball, the probability that it came from the box with the red and green ball was lower than the probability that it came from the box with two green balls. If X selected a red ball, the probability would have been lower than the probability that the ball came from the third box. The actual probability that the remaining ball in the box is red is .3333.

The mathematical concept behind this example of a counter intuitive probability is that of conditional probability. In conditional probabilities one is calculating the probability of an event given the assumption or fact that another event has occurred. Consider another similar example: assume that the passenger sitting next to you on an airplane flight tells you that he has two children. He then describes one of his children as his daughter. What is the probability that his other child is a girl? Again, one's first intuition is that the other child is either a boy or a girl so the probability of both children being girls would be .5. When one reviews all of the possible outcomes, however, a different answer emerges. Four possible outcomes, listed in order of birth, exist for gender distribution among children: boy-boy, boy-girl, girl-boy, and girl-girl. In three of these four cases, one of the children is a girl. Accordingly, given the knowledge that there is at least one girl, the probability of a girl-girl outcome is .333. If the passenger had informed you that his eldest child was a girl, the probability would change to .5 because only the last two of the listed outcomes would be possible.[76] Thomas Bayes set forth the following theorem for computing conditional probabilities: $P(A/C) = P(A\&C)$ divided by $P(C)$, where A and C are events with probabilities, $P(A)$ and $P(C)$, and C is an event which has occurred. In applying Bayes theorem to the first portion of the example, C would equal at least one child is a girl and A would equal the event that the other child is a girl. $P(C)$ would equal .75 and $P(A\&C)$ would equal .25. .75 divided by .25 would equal .333, the probability that the other child is a girl.

Pierre Lecomte du Noüy explained that the calculus of the probability of an event is equal to the number of outcomes favorable to the event divided by the total number of possibilities with all possible outcomes considered to be equally probable. If there are two possible outcomes, such as in the tossing of a symmetrical thin coin to see if the coin comes up heads or tails, the total number of possibilities is 2, with heads or tails equally probable. The probability for heads is 1 divided by 2 or .5. If we are dividing lots with 10 different possible lot outcomes, the probability of one particular outcome will be .1.

In the calculus of the probability of an event which requires two outcomes to appear in sequence, the probability is determined by multiplying the probability of the first outcome by the probability of the second outcome. For example, the probability of throwing two consecutive 7s in a game of dice is equal to the product of the ratio of the first favorable outcome (1/6) multiplied by the ratio of the second favorable outcome (1/6) or 1/36 which equals a probability of .0277. The probability of throwing 10 consecutive 7s is 1/60, 466, 176 (rounded) or .000,000,016.[77]

As discussed above, DNA is found in all living matter, and the genetic information in DNA functions in many respects in a manner similar to the letters of the Roman alphabet of the English language. Just as the sequence of alphabet letters determine

the meaning or information conveyed by the letters, the sequence of nucleotides and amino acids determine the meaning of
the genetic message. The DNA "alphabet" conveys instructions in
a manner somewhat similar to the Roman alphabet. Consequently, a calculation of the probability of monkeys typing Shakespeare may be interesting in considering an analogy, with some
limitations as discussed below, to the possibility of an unguided,
random generation of the simplest form of life. Thomas Huxley
asserted that a large number of monkeys typing randomly on a
large number of typewriters would eventually type the complete
works of Shakespeare. The assertion is based on the probability
theorem of the Strong Law of Large Numbers. Huxley wrote in a
time when the universe was considered by many scientists to be
in a steady state with an infinite age. Theoretically, the example
may work, but the universe is not old enough and the number of
monkeys available are insufficient to allow for the typing of even
a short passage of Shakespeare. Time is not the hero of the plot;
the monkeys do not have sufficient time.

What are the odds of an accidental typing of Shakespeare?
Consider the following very brief passage from *Macbeth* containing 379 letters, each one selected from our alphabet of 26:

> She should have died hereafter;
> There would have been a time for such a word.
> To-morrow, and to-morrow, and to-morrow
> Creeps in this petty pace from day to day
> To the last syllable of recorded time;
> And all our yesterdays have lighted fools
> The way to dusty death. Out, out, brief candle!
> Life's but a walking shadow, a poor player,
> That struts and frets his hour upon the stage
> And then is heard no more. It is a tale
> Told by an idiot, full of sound and fury,
> Signifying nothing.[78]

Ignoring the space between words and lines, the probability
of producing this language, not all of *Macbeth* but just this very
small portion, is 26^{379}. By the following equations we can translate
that number into a power of 10:

$$10^x = 26^{379}$$
$$\log 10^x = \log 26^{379}$$
$$x \log 10 = 379 \log 26$$
$$\log 26 = 1.414973348$$
$$x = 536.275$$

Accordingly, the probability is one in 10^{536}. One chance in 10^{536} is more than extremely unlikely. To put that number in context there are only 10^{80} atoms in the known universe. As we will discuss in a later section concerning precision in physics, Paul Davies equates the odds of one chance in 10^{60} to hitting a one inch target with the random shot of a bullet from a distance of twenty billion light years! As noted above, mathematicians normally regard anything with a probability of less than one in 10^{50} as mathematical impossibility.

Assuming that the Big Bang occurred 15 billion years ago and that one million monkeys started typing at Planck time (10^{-43} of the first second) and that each monkey types one letter every second, over a million billion years would be required to produce all possible combinations. To put time in terms of a power of 10, only 10^{18} seconds have occurred in all of time. As with the time available for abiogenesis, the monkeys simply do not have sufficient time in 10^{18} seconds to have any real chance of typing this short passage from Shakespeare.[79] When we turn to calculations of mathematical probabilities for the unguided, random development of life, we find odds that are even more remote, especially given the finite time limit of 130 million years.

Before exploring other calculations, we will review an example of a calculation which does not validly compute the mathematical probabilities of unguided, random events. In a popular science reference book, accurate mathematics, coupled with unwarranted assumptions, produce the following questionable analysis:

> To show you how efficient natural selection can be, imagine that you want to have the entire Bible typed by a wild monkey. What are the chances that such a monkey, typing at random, will come up with the Bible neatly typed without a single error? The English

Bible (King James translation) contains about 6 million letters. The chances of success, therefore, are 1/$26^{6,000,000}$, as there are 26 letters in the English alphabet. This is equal to $10^{-8,489,840}$. I wouldn't exactly wait around. Suppose, however, that I introduce a control (the environment) that wipes out any wrong letter the monkey may type. Typing away at one letter per second and assuming an average number of 13 errors per letter (half of 26), the monkey will produce the Bible in 13 x 6,000,000 seconds = 2.5 years. . . . This is precisely what the environment does. It knows what kind of organism would best fit and if the wrong one appears it rejects it as you reject a wrong letter. All the environment does is to effectively eliminate all the random changes that are in the wrong direction. Given the chemical and environmental conditions of the primitive earth, the appearance of life was a foregone conclusion.[80]

Remembering the section in this book on valid and invalid reasoning processes, the unwarranted and unproven assumptions contained in this analysis are remarkable. An invalid assumption is that the "environment wipes out any wrong letter," because this is the very assumption which must be proved to show that random processes can produce the Bible. Note that, without any evidence, the term "environment" is endowed with characteristics including powers of intelligence to "know" and "reject" wrong letters. This is an example of *circulus in probando*; the answer is assumed in the premise that the "environment" will "know what kind of organism is best and reject wrong letters or sequences." The assumptions are made without any rationale and ignore a fundamental principle in science: *natural selection does not exist in prebiological molecules.* It is generally agreed that natural selection can only act on systems capable of replication. Natural selection alone is not sufficient to explain the origin of life. The relevant analogy is to the origin, not to the replication or mutation of life. The analogy fails because it does not relate to comparable terms in a consistent context.

The fallacy in this circularity is compounded by the arbitrary selection of the number 13 with an assumption that only one half of the Roman alphabet will be needed to produce the correct letter. What is the empirical evidence or rationale giving any validity to the assumption of 13 errors per letter? If a golfer assumes that she will birdie nine out of eighteen holes in golf and arbitrarily assigns a score of one under par for the last nine holes and plays only the first nine holes with a bogey on each hole, she will have assigned herself a score for the eighteen holes equal to par. But the reasonableness of that score is based on an assumption which must have some relationship to her abilities and to the probability of her scoring nine consecutive birdies. The empirical results of her previous scores provide a rational basis for testing the validity of her assumptions.

The analysis also uses the term "environment" as the entity responsible for appointing the precise desirable values or conditions required to achieve a particular purpose. When the term "environment" is used with characteristics similar to a conscious mind, the question arises concerning the distinction between the term and the word "intelligence" or "Superior Intelligence". If one uses the term in a manner implying intelligence, one is no longer discussing random, chance or accidental processes. The monkey is not producing a document by chance under the conditions given in the quotation.

Richard Dawkins constructs a similar failed analogy in his book *The Blind Watchmaker*. Dawkins understands the odds against chance as the sole cause of life and presupposes that the process of natural selection determines the "correct" letters which the monkey preserves.[81] However, for the monkey to preserve the correct letters in the sequence requires an assumed intelligence apart from and greater than the intelligence of the monkey. This intelligence must have knowledge of the letters which construct a meaningful sentence. Without such an intelligence, no principle exists for deciding which letters should be preserved. Natural selection does not qualify as such an intelligence, because it is a process, not something like an intelligent mind which knows the alphabet and the structure of a meaningful sentence. Dawkins cannot have it both ways. He cannot logically assert that a process without the characteristics of a mind has the characteristics of a mind and the knowledge required to "know" which letters to preserve. Such an assertion fails because it assumes a self-contradiction. *Cadit quaestio*.[82]

Although useful in apprehending random probability, word game models have some limitations as analogies to the messages of the genetic code. In all of the above calculations, the analogy between alphabet letters and the origin of DNA, RNA, and protein processes is not exact. The odds against random models are even higher in the origin of these processes than in word formation. There are no intersymbol influences among amino acids polymerized in proteins, but such influences exist between letters. These influences include the rules of composition, grammar, spelling and the frequencies of use of certain letters in English. Also, the complexity of proteins is far greater than the complexity of the English language. The computational complexity of proteins is grossly underestimated in any word game analogy; the odds against abiogenesis in the physical world is much greater.[83]

3.7. Mathematical probability of random protein/ enzyme and bacterium formation

3.7.1. Calculations of Sir Fred Hoyle and Chandra Wickramasinghe for random generation of a simple enzyme and calculations for a single celled bacterium

Sir Fred Hoyle and Chandra Wickramasinghe understood that any of the simplest living cells such as bacteria was extremely complex, containing many nucleic acids and enzymes and molecules, all comprised of thousands of atoms, all joined together in a precise sequence. Although he is an evolutionist (but not a Darwinist) and an atheist, Hoyle sees the mathematical statistical difficulty. In his calculations of the probability of life emerging from the chance interactions of chemicals, Hoyle assumed that the first living cell was much simpler than today's bacteria. However, his calculation for the likelihood of even one very simple *enzyme* arising at the right time in the right place was only one chance in 10^{20} or 1 in 100,000,000,000,000,000,000.

Because there are thousands of different enzymes with different functions, to produce the simplest living *cell*, Hoyle calculated that about 2,000 enzymes were needed with each one performing a specific task to form a single bacterium like *E. coli*. Computing the probability of all of these different enzymes forming in one place at one time to produce a single bacterium, Hoyle and his colleague, Chandra Wickramasinghe, calculated the odds at 1 in $10^{40,000}$. This number is so vast that any mathematician would agree that it amounts to total impossibility. As noted above, the total atoms in the observable universe are estimated to be only approximately 10^{80}.

Hoyle and Wickramasinghe concluded that life could not have appeared by earthbound random processes even if the whole universe consisted of primeval soup. The enormous information content of even the simplest living system cannot be generated by accidental processes. Any theory with a probability of being accurate larger than 1 in $10^{40,000}$ must be considered superior to random processes. The probability that life was assembled by an intelligence has a vastly greater probability.[84] They argue that life could not have emerged on earth from unguided, random processes:

> No matter how large the environment one considers, life cannot have had a random beginning . . . there are about two thousand enzymes, and the chance of obtaining them all in a random trial is only one part in $(10^{20})^{2000} = 10^{40,000}$, an outrageously small probability that could not be faced even if the whole universe consisted of organic soup. If one is not prejudiced either by social beliefs or by a scientific training into the conviction that life originated on the Earth, this simple calculation wipes the idea entirely out of court . . . the enormous information content of even the simplest living systems . . . cannot in our view be generated by what are often called "natural" processes, . . . For life to have originated on the Earth it would be necessary that quite explicit instruction should have been provided for its assembly . . . There is no way in which we can expect to avoid the need for information, no way in which we can simply get by with a bigger and better organic soup, as we ourselves hoped might be possible a year or two ago."[85]

Chandra Wickramasinghe adds the following dramatic summary statement:

> The chances that life just occurred are about as un-
> likely as a typhoon blowing through a junkyard and
> constructing a Boeing 747.[86]

Francis Crick, who with James Watson shared the Nobel prize for their double helical model for deoxyribonucleic acid (DNA), knows that life appeared on earth almost as soon as the planet had cooled to a point compatible with life. He also understands the mathematical impossibility of random development of life on earth. Crick, Fred Hoyle, Svante Arrhenius, Leslie Orgel, and Thomas Gold are among the scientists who have turned to the panspermia hypothesis to explain life on earth. These scientists are aware of the enormous odds against abiogenesis on earth and of the extremely short period of time available on earth for life to form by chance. They conclude that life on earth cannot be explained by chance, and some of them hypothesize that the seeds of life were sent to earth in a spaceship from a dying planet or are being dispersed all over the universe by some unexplained natural processes.[87]

Harold Morowitz notes that all approaches to a search for a plausible theory of life's origin are necessarily influenced by metaphysical philosophical perspectives. He rejects the extraterrestrial seeding hypothesis because it violates Ockham's razor which states that unnecessary assumptions should be avoided in the construction of hypotheses: *"Non sunt entia multiplicanda practer necessitatem."*[88] Moreover, the panspermia hypothesis does not really advance the cause of proponents of abiogenesis. The calculations in this book demonstrate that accidental abiogenesis is mathematically impossible even if one considers the universe to be as old as 15 billion years.

3.7.2. Calculations of Hubert Yockey for random generation of a single molecule of iso-1-cytochrome c protein

Perhaps a more accurate and improved calculation was made by Hubert P. Yockey, the preeminent authority on information theory and biology, who calculated the mathematical probability of life emerging by chance from a prebiotic soup and came to the conclusion that Hoyle was too optimistic! Yockey noted that because all amino acids are not equally probable, a correct calculation cannot simply multiply the number of functionally equal amino acids at each site to arrive at the number of sequences. Yockey selected iso-l-cytochrome c as one model protein with known functionally equivalent amino acids and proceeded to calculate the probability of the generation of one single molecule of that specific protein.

He assigned the responsibility of amino acid selection and their polymerizing to form proteins to three Fates, acting as *dei ex machina* in a Greek drama. Lachesis was the caster of 110 icosahedral dice; Clotho, the spinner of the thread of life, polymerized them; and Atropos cut the thread when Lachesis assigned an amino acid to a non functionally equivalent site. Yockey asked the question: what is the probability that Lachesis and Clotho will build a chain of 110 amino acids of the iso-l-cytochrome c without Atropos cutting it?

The probability calculated was 2×10^{-44}. Yockey then noted that the realistic odds are much worse because even among all proponents of the prebiotic soup, there is agreement that many non-proteinous amino acids and analogues existed in the soup with the proteinous amino acids. He summarized his analysis as follows:

> Let us remind ourselves that we have calculated the probability of the generation of only a single molecule of iso-1-cytochrome c. Of course, very many copies of each molecule must be generated to form the protobiont. . . . I am using probability as a measure of degree of belief. It is clear that the belief that a molecule of iso-1-cytochrome c or any other protein could appear by chance is based on faith. And so we see that even if we believe that the "building blocks" are available, they do not spontaneously make proteins, at least not by chance. The origin of life by chance in a primeval soup is impossible in probability in the same way that a perpetual motion machine is impossible in probability. The extremely small probabilities calcu-

lated in this chapter are not discouraging to true be-
lievers . . . or to people who live in a universe of infinite extension that has no beginning or end in time.
In such a universe all things not *streng verboten* will
happen. In fact we live in a small, young universe
generated by an enormous hydrogen bomb explosion
some time between 10×10^9 and 20×10^9 years ago. A
practical person must conclude that life didn't hap-
pen by chance.[89]

3.7.3. Calculations of Bradley and Thaxton for random production of a single protein

Walter L. Bradley and Charles B. Thaxton calculated the probabil-
ity of a random formation of amino acids into a protein to be 4.9×10^{-191}. They began with the assumption that the probability of
starting with an L-amino acid was .5, and the probability of two
L-amino acids joining with a peptide bond was also .5. They assumed that the twenty necessary amino acids existed in equal
concentration in the prebiotic soup so that the probability of the
right amino acid in the required position was .05.

Bradley and Thaxton were also generous towards the proponents of random processes when they also assumed that all of the
chemical reactions would be with amino acids, ignoring the high
probability of reactions with non-amino acid chemicals. They calculated the probability of the necessary placement of one amino
acid to be $.5 \times .5 \times .05$ or .0125. This, of course, meant that the
probability of assembling N such amino acids would be .0125 x
.0125 for N terms. Assuming a protein with 100 amino acids (.0125
x .0125 for 100 terms), the mathematically impossible probability
would be 4.9×10^{-191}.

Bradley and Thaxton noted their agreement with Hubert P.
Yockey and concluded that even assuming that all the carbon on
earth existed in the form of amino acids and reacted at the greatest possible rate of 10^{12}/s for one billion years (when actually only
130 million years were available), the mathematically impossible
probability for the formation of one functional protein would be
$\sim 10^{-65}$.[90]

3.7.4. Calculations of Harold Morowitz for single celled bacterium developing from accidental or chance processes

The difficulties in producing a protein from the mythical prebiotic soup are very large, but more difficult still is the probability of random processes producing the simplest living cell which represents an overwhelming increase in complexity. Harold Morowitz calculated the probability of broken chemical bonds in a single celled bacterium reassembling under ideal chemical conditions. He assumed that only constructive chemical processes were acting (under natural conditions 50 percent of chemical processes are destructive) and that all of the amino acids were bioactive (in a natural environment 75 percent of amino acids are not bioactive). Morowitz computed the odds against the cell reassembling to be one in $10^{100,000,000,000}$. He summarized his computation:

> . . . no amount of ordinary manipulation or arguing about the age of the universe or the size of the system can suffice to make it plausible that such a fluctuation would have occurred in an equilibrium system. It is always possible to argue that any unique event would have occurred. This is outside the range of probabilistic considerations, and really, outside of science. We may sum up stating that on energy considerations alone, the possibility of a living cell occurring in an equilibrium ensemble is vanishingly small. It is important to reiterate this point as a number of authors on the origin of life have missed the significance of vanishingly small probabilities. They have assumed that the final probability will be reasonably large by virtue of the size and age of the system. The previous paragraph shows that this is not so: calculable values of the probability of spontaneous origin are so low that the final probabilities are still vanishingly small.[91]

Morowitz also calculated the increase in chemical bonding energy required in forming an *E. coli* bacterium and the probability of such a bacterium forming spontaneously anywhere in the

entire universe over a period of five billion years under equilibrium conditions. In computing the odds to be one in $10^{10^{(110)}}$,
Morowitz wrote:

> What is very clear . . . is that if equilibrium processes
> alone were at work, the largest possible fluctuation in
> the history of the universe is likely to have been no
> larger than a small peptide. Again, we stress in a very
> firm quantitative way, the impossibility of life origi-
> nating as a fluctuation in an equilibrium ensemble.[92]

3.7.5. Calculations of Bernd-Olaf Küppers for the random generation of the sequence of a bacterium

Proceeding from the realistic assumption that all sequence alternatives of a nucleic-acid molecule are physically equivalent,
Bernd-Olaf Küppers concluded that the unguided, random formation of a predefined sequence (such as the specific sequence of
the nucleotides in the DNA molecule) is reciprocally proportional
to the number of all combinations of possible sequences. Küppers
noted that Michael Polanyi correctly emphasized that if the reverse assumption were true and the sequence of a nucleic-acid
molecule was predetermined by chemical bonds, then a nucleic-
acid molecule would not have the capability to store information
necessary to replicate living matter. (See further discussion on
Polanyi in section 4.1.3.4.). In calculating the expectation probability for the nucleotide sequence of a bacterium, Küppers demonstrated the reason mathematicians have severe problems in
accepting the assumptions of random origins:

> The human genome consists of about 10^9 nucleotides,
> and the number of combinatorially possible sequences
> attains the unimaginable size of $4^{1000\,\text{million}} \approx 10^{600\,\text{million}}$.
> Even in the simple case of a bacterium, the genome
> consists of some 4.10^6 nucleotides, and the number of
> combinatorially possible sequences is $4^{4\,\text{million}} \approx 10^{2.4\,\text{million}}$.
> The expectation probability for the nucleotide se-
> quence of a bacterium is thus so slight that not even

the entire space of the universe would be enough to make the random synthesis of a bacterial genome probable. For example, the entire mass of the universe, expressed as a multiple of the mass of the hydrogen atom, amounts to about 10^{80} units. Even if all the matter in space consisted of DNA molecules of the structural complexity of the bacterial genome, with random sequences, then the chances of finding among them a bacterial genome or something resembling one would still be completely negligible.[93]

3.8. Additional challenges from complexity

The case against chance or accident is compelling. Living matter at the simplest level is exceedingly intricate. Discoveries in molecular biology disclose a world of staggering complexity. Even a single celled bacterium is comprised of ten million million atoms and an enormous amount of instructions or information content. Michael Denton summarizes the perspectives of many mathematicians and biologists who are skeptical about the formation of life by accidental, random processes:

> At the Wistar Institute Symposium in 1966 (entitled, "Mathematical Challenges to the Darwinian Interpretation of Evolution") which brought together mathematicians and biologists of impeccable academic credentials, Sir Peter Medawar acknowledged in his introductory address the existence of a widespread feeling of skepticism over the role of chance . . . Perhaps in no other area of modern biology is the challenge posed by the extreme complexity and ingenuity of biological adaptations more apparent than in the fascinating new molecular world of the cell . . . To grasp the reality of life as it has been revealed by molecular biology, we must magnify a cell a thousand million times until it is twenty kilometers in diameter and resembles a giant airship large

enough to cover a great city like London or New
York. What we would then see would be an object of
unparalleled complexity and adaptive design. On the
surface of the cell we would see millions of openings,
like the port holes of a vast space ship, opening and
closing to allow a continual stream of materials to
flow in and out. If we were to enter one of these
openings we would find ourselves in a world of supreme technology and bewildering complexity . . . It
is the sheer universality of perfection, the fact that
everywhere we look, to whatever depth we look, we
find an elegance and ingenuity of an absolutely transcending quality, which so mitigates against the idea
of chance. Is it really credible that random processes
could have constructed a reality, the smallest element
of which—a functional protein or gene—is complex
beyond our own creative capacities, a reality which is
the very antithesis of chance, which excels in every
sense anything produced by the intelligence of
man?[94] (parentheses added)

For the proponents of accident another daunting obstacle is
the explanation of the incredible increase of complexity that one
finds in the human brain. The vast complexity of a single cell
pales in comparison to the complexity of the human brain which
consists of more than ten thousand million nerve cells with each
cell containing ten thousand to one hundred thousand fibers
connecting the brain cells so that the total connections among
brain cells total a thousand million million or 10^{15}. This is an incredibly large number, especially when one considers that each
brain fiber provides a special function in the brain's communication system. To put the number in perspective these connections
represent over 100 times the number of connections in the total
network of communications on the planet earth! The probability
of the assembly of such a system even by intelligent human beings is exceedingly small. The argument that such an assembly
was performed by accident stretches credulity. *Non ex quovis ligno
Mercurius fit.*[95] To quote Michael Denton again: "Because of the

vast number of unique adaptive connections (required) to assemble an object remotely resembling the brain, (such an assemblage) would take an eternity even applying the most sophisticated engineering techniques. Undoubtedly, the complexity of biological systems in terms of the sheer number of unique components is very impressive; and it raises the obvious question: could any sort of purely random process ever have assembled such systems in the time available?"[96]

PART IV

THE PROBLEM OF COMPLEXITY: THE GENERATION OF SUFFICIENT INFORMATION CONTENT

4.1. Absence of plausible method of generating sufficient information content into inorganic matter even in a system far from equilibrium

4.1.1. Insufficiency of energy flow alone to generate adequate information content

Systems near equilibrium are simply not capable of producing the complexity required for life. The Second Law of Thermodynamics states that in any spontaneous process in such a system there is an increase in disorder or entropy. Systems near equilibrium will always move toward disorder or entropy. The Second Law is time's arrow which points in the direction of equilibrium so that in any spontaneous change, the amount of energy available (free energy) decreases and the randomness increases, i.e., the more time available, the greater the entropy or disorder. Life in these systems could not have developed by chance processes.

The probabilities of abiogenesis appear greater when considering an open system with an energy source maintaining the system far from equilibrium and from the disorder which inexorably occurs pursuant to the Second Law in equilibrium processes. Although the earth has an energy source from the sun, energy alone is not sufficient to support abiogenesis. Dynamite can be a source of energy, but unless the energy from its explosion is directed in an intelligent manner, its energy will be more destructive than constructive. For abiogenesis to occur, energy flow must be joined to a mechanism which will direct it to generate sufficient information content into inert matter.[97] Information content is the minimum number of instructions needed to specify the structure. The information content of living systems contains an enormous amount of specified instructions. The complexity existing in this information content is the principal characteristic or the *sine qua non* of living matter.

In reviewing the effects of energy flow one must distinguish between the maintenance of order in a living system and the origination of a living system from inert matter. Energy flow simply maintaining a system far from equilibrium and protecting it from the effects of the Second Law may sustain the order in a system, but energy flow alone is not sufficient to explain the complexity of life's origin. For example, Toby, my family's golden retriever, eats heated frozen, pre-packaged turkey dinners to provide himself with energy which builds and maintains his body. To maintain his life, he needs to have a stomach, liver, and intestines which provide a mechanism to join the energy available from the turkey dinner to the work required to sustain his body. This example of the maintenance of a golden retriever's body is fairly simple to understand because the energy flow is joined to the required work by the dog's mechanism of DNA, enzymes, and RNA. The origin of this mechanism, however, is a deep, unsolved mystery.[98] The solution to the puzzle of life's origin requires an explanation of the development of molecules with intense information content. By what means is the energy flow which keeps a system far from equilibrium capable of generating information content? How did the mechanism which stores, transfers, and directs information arise spontaneously? Natural selection is not a viable explanation for the origin of DNA and enzymes, because, as noted above in the critique of the mathematical probabilities of a monkey typing the Bible, natural selection only acts within systems which already have replicating capacity. *Again, natural selection does not exist in prebiological molecules.*

These points are worth restating: Energy flow from a source, like the sun, can keep a system far from equilibrium. However, the energy flow which maintains a system far from equilibrium does not contribute towards the origin of life if the energy flow is not directed in some manner into generating information content into inorganic matter. The energy flow does remove the system from equilibrium and prevent the total disorder which flows from the Second Law, but that alone is not sufficient to begin life, because life requires energy flow *to be directed* to produce information content in inert matter.[99] The issue of the generation of information content is the fundamental problem in the origin of life theories, not the influence of the Second Law of Thermodynamics.

As discussed many times in this book, scientists frequently confuse the concepts of order and complexity. To construct a plausible theory for the origin of life, scientists need to discover a theory which explains the generation of complexity, not the generation of order. The Second Law of Thermodynamics addresses the orderliness of energy. Order may arise spontaneously in inorganic systems far from equilibrium. In terms of the formation of the first living organism, however, the applicability of the Second Law in a system far from equilibrium is not so significant, because complexity rather than order is the issue. In this sense order is *nihil ad rem*.[100]

Complexity depends upon a structure's information content which is the minimum number of instructions necessary to specify the structure. The speculations of the leading theorists concerning the generation of order in a system removed far from equilibrium (such as Prigogine, Cairns-Smith, Wächtershäuser, Morowitz, and Kauffman discussed below) fail because they describe a scenario for the formation of order rather than for the generation of complexity. The real issue for a plausible scenario for the origin of life is the generation of complexity reflected in the genetic message of the genome. The stumbling block remains the genetic code as found in the RNA, DNA, and enzyme process. The consistent failure to synthesize protein or DNA reflects the problem of finding a means of storing and transferring the information and instructions required for life.

4.1.2. *The improbability of RNA as a catalyst for the origin of life*

RNA is able to function as a protein to some extent and act as a catalyst to join two amino acids with a peptide bond or as a molecule capable of making the bonds that join amino acids to RNA, but RNA catalysis is not as versatile as that of protein enzymes. More importantly, if one prebiotic molecule could perform all the tasks of proteins, DNA, and RNA, the complexity of that molecule would have to equal the sum of the complexity of DNA, RNA, and proteins. The odds against the formation of such an unusually talented molecule are no less than the spontaneous and simultaneous formation of DNA, RNA, and proteins. The information contained in this super molecule would have to equal the information contained in all three kinds of molecules.

RNA has ribose as its five carbon sugar. Ribose is only one of a variety of sugars (DNA has deoxyribose as its sugar) and is never the primary product. In the prebiotic environment only one synthesis of ribose is plausible: the polymerization of formaldehyde. The proposal of an RNA catalyst for the origin of life raises many questions which are as difficult as the one the proposal attempts to solve. Michael J. Behe notes the difficulty in this proposal:

> The big problem is that each nucleotide "building block" is itself built up from several components, and the processes that form the components are chemically incompatible. Although a chemist can make nucleotides with ease in a laboratory by synthesizing the components separately, purifying them, and then recombining the components to react with each other, undirected chemical reactions overwhelmingly produce undesired products and shapeless goop on the bottom of the test tube. Gerald Joyce and Leslie Orgel—two scientists who have worked long and hard on the origin of life problem—call RNA "the prebiotic chemist's nightmare."[101]

Even if one accepts the idea that a self-replicating RNA molecule emerged from a prebiotic soup, such a molecule could not serve as a catalyst for life unless it was a very unique form of RNA which contained an unusual chemistry unlike the chemistry of most RNA, which does not contain the required catalytic characteristics that would be effective in any origin of life scenario. Behe quotes origin of life scientists Joyce and Orgel as they set forth their quandry:

> This discussion . . . has, in a sense, focused on a straw man: the myth of a self-replicating RNA molecule that arose de novo from a soup of random polynucleotides. Not only is such a notion unrealistic in light of

our current understanding of prebiotic chemistry, but it should strain the credulity of even an optimist's view of RNA's catalytic potential. . . . Without evolution it appears unlikely that a self-replicating ribozyme could arise, but without some form of self-replication there is no way to conduct an evolutionary search for the first, primitive self-replicating ribozyme.[102]

RNA is a molecule of such substantial complexity that some scientists have hypothesized that a simpler genetic system preceded the development of RNA. Yockey notes that this hypothesis, even if true, does not solve the puzzle of the generation of information content into inorganic matter: "This suggestion proposes to move the problem a step nearer to the protobiont but still encounters the primary questions of the generation of the genetic message and of the genetic code between the alphabet of the genetic sequence that stores and replicates the genetic messages and the alphabet of the protein sequences that implement function."[103]

Robert A. Shapiro, a DNA chemist at New York University, has demonstrated that synthesis of ribose and deoxyribose sugar under plausible prebiotic conditions was impossible.[104] RNA is difficult to synthesize under the best of conditions, much less under plausible prebiotic ones. The trend among the opinions of scientists today is that the random generation of RNA in plausible prebiotic conditions was extremely unlikely.

4.1.3. Other theories of self-organization in nonequilibrium systems

As mentioned above, some scientists who reject the chance or accidental origin of life scenarios have proposed several theories where the emergence of life results from the laws of physics and chemistry. Under these theories life is the consequence of the properties of matter and determined from the inception of the universe. At the present time there is no scientifically plausible origin of life scenario among the various self-organization scenarios in the offering. We shall consider several of the more prominent theories. All of the following theories reject the emergence of life by chance or accident alone, and all of the theories are extremely speculative.

For readers with theistic concerns, it may be interesting to note that many of the theorists do not view deterministic self-organization scenarios as antitheistic. For example, having rejected the prebiotic soup origin of life by chance scenario and proposed a rather novel theory whereby metabolism recapitulates biogenesis with lipid vesicles acting as precursors to life, Harold Morowitz emphasizes the consistency of his theory with a religious perspective:

> It is widely assumed that to hold the idea that life began by deterministic processes on the surface of the earth is, of necessity, an anti-religious point of view. This is certainly not the case, as this view is acceptable to Buddhists, Taoists, liberal Protestants, Reform Jews, Roman Catholics, and pantheists. Indeed, the idea of life as necessary would seem, at the minimum, to presuppose that the universe is infused with something like a creative intelligence. Although the approach may not tell us much about purpose, it is certainly consistent with the feeling of awe and wonderment that is an existential point of much of contemporary religious thought.[105]

Whether Morowitz is contemplating a deist theology is not an issue for this book. A detailed discussion of theological terms is not within the scope of the questions presented. Such a discussion should be addressed elsewhere. For purposes of answering the second question presented we will emphasize that all of the different self-organization theories fail because they do not present a plausible method of generating sufficient information content in the time available. It is worth repeating that these theories confuse the concepts of complexity and order which in many respects are opposites. A protein contains little order but high complexity (informational or instructional content). Although the term "complexity" is frequently used in these scenarios, what is meant is not information content as the term is used in this book, but organization or order. Again, information content means the minimum number of instructions necessary to specify a structure. When a self-organization theorist writes that a system is complex with a high degree of order, he or she frequently means that the system is organized to a high level. This is not the meaning of the term "complexity" in information theory. Hubert Yockey calculated that the information content of the cytochrome c genetic message was between 233 and 374 bits to record the information instructions in one molecule of iso-1-cytochrome c.[106] The paradigms for the emergence of life are algorithms which must contain at least as much information content as the genetic messages they claim to generate. The information content in iso-1-cytochrome c is much greater than the information content in the paradigms of self-organization scenarios presently in the offering. All of these scenarios fail to give a plausible explanation which can meet the criteria imposed by information theory.[107] *Alia tendanda via est.*[108]

4.1.3.1. Order without specified complexity

Nobel laureate Ilya Prigogine emphasizes that highly ordered behavior in inorganic matter may appear spontaneously in systems far from equilibrium. The order noted in Prigogine's writings, however, is not very useful in resolving the enigma of the origin of life. Prigogine fails to distinguish between order and specified complexity. One of his examples of a spontaneous ordering is the vortex formed by water molecules as water is influenced by the force of gravity and flows down a bathtub drain. Prigogine's example is an ordered structure with low information content. An ordered structure may have low or high information content. The ordered movement of water in the vortex has low information content and is not at all analogous to the irregular structure of living systems which contain vast amounts of information. A genetic message is conveyed by an aperiodic structure with high information content. Again, order is not synonymous with complexity. A structure's specified complexity relates to the quantity of its information content with high complexity requiring more information content which requires more instructions necessary to specify the structure. Highly complex structures require many instructions. A structure may be highly ordered, such as a crystal, but contain very few instructions. Applying the analogy of language, *The Oxford History of the American People* has a high information content, but a series of one thousand pages with only the letters ABC appearing in repeating order on each line has a higher level of order and a very low level of information content.

Except for written language or human artifacts, no inorganic matter has specified complexity. The fundamental distinction between living systems and inorganic matter is this specified complexity, not simple, periodic order. In DNA the nucleotide sequence is highly irregular and aperiodic, similar to letters in a written communication. A crystal, on the other hand, has a simple, periodic repetitive order with very few instructions required to specify its structure. The generation of information content is the central question in the search for a plausible, scientific scenario for the origin of life.

Hubert Yockey argues that Prigogine's application of non-equilibrium thermodynamics to biology is inappropriate because it confuses the distinction between order and complexity. Prigogine's theory fails because it concerns the generation of order, not the generation of complexity. A plausible theory for the origin of life must address the question of the genetic message and the generation of information content.[109] Thaxton and his colleagues agree with Yockey. They comment on the lack of a connection between Prigogine's concept of order in systems far from equilibrium and the work required to generate information rich macromolecules similar to the ones found in DNA:

> Regularity or order cannot serve to store the large amount of information required by living systems. A highly irregular, but specified, structure is required rather than an ordered structure. This is a serious flaw in the analogy offered. There is no apparent connnection between the kind of spontaneous ordering that occurs from energy flow through such systems and the work required to build aperiodic information-intensive macromolecules like DNA and protein. Prigogine et. al. suggest that the energy flow through the system decreases the system's entropy,

leading potentially to the highly organized structure of DNA and protein. Yet they offer no suggestion as to how the decrease in thermal entropy from energy flow through the system could be coupled to do the configurational entropy work required.[110]

4.1.3.2. Silicate crystals: self-replication without specified complexity

A. G. Cairns-Smith, a biochemist at the University of Glasglow, Scotland, hypothesizes that clays may have formed the first self-replicating structures. Cairns-Smith devised an elaborate theory which proposed that amino acids were concentrated by adsorption on clay. Cairns-Smith reasoned that because clay acts as an industrial catalyst, it served as a primitive catalyst in encouraging flawed crystals to form information content in carbon-chained molecules. He rejected the concept of a prebiotic soup and proposed that the first living organism resulted from the growth of one crystal on the surface of the lattice of another crystal. He called his theory of replicating clays the genetic takeover and proposed RNA as the takeover molecule.[111]

Cairns-Smith noted that the microcrystals of clay consist of a regular silicate lattice with a routine pattern of ionic locations but with deviated distribution of metals at those locations. He regarded the metal ions as carriers of information similar to the nucleotide basis in an RNA molecule. These ions can form irregular patterns of electrostatic potential which can adsorb molecules to the surfaces, and, as hypothesized by Cairns-Smith, perform the same function as RNA in a crude fashion. According to his theory, one day crystal discovered that RNA is a better genetic substance than clay, and RNA was formed.

Cairns-Smith's theory falters in failing to explain complexity and in failing to distinguish between order and complexity. Again, complexity is the *sine qua non* of living matter. The distinction between living and non-living structures is in their complexity which is represented by the high information content found in living organisms. No experiment has produced anything like this complexity. Crystals are not a viable explanation for the origin of a mechanism which would generate sufficient information content into inert matter to produce the genome necessary for life. Although crystals are ordered with periodic arrangements of atoms, they carry very little information.

Nucleic acids and proteins are information macromolecules with aperiodic structures arranged in a specified sequence. The specified sequence of the base sequence of a DNA molecule has an unpredictable pattern with flexibility which allows the conveyance of a vast amount of information. If DNA consisted of the same type of order as a crystal, it would only be capable of repeating a simple message over and over again.[112] DNA represents an entirely different type of order than the type of order found in a crystal. Highly ordered crystals are repetitive in structure. They are similar to the old story of a law student who had too little sleep and too much caffeine and wrote the same sentence over and over again on every line of his examination book. His examination essay was very ordered and very redundant. Redundancy is the main characteristic of crystal structure, but complex sequences and information are characteristics of life forms.

The distinction between order and complexity is well delineated in information theory which emphasizes the quantification and measurement of information content. A crystal has a highly ordered structure but low intelligence or information content. The DNA molecule has a high information content with a complicated set of instructions for the assembly of the organism. It has complexity. Crystal structures may be highly ordered, but have a low information content and do not have complexity.

In regarding the crystal imperfections as the source of the RNA, DNA and enzyme system, Cairns-Smith is "grossly mistaken" in his hypothesis that the information density in a crystallite is at all similar to the information content in DNA.[113] *Ex arena funiculum nectis.*[114] A large chasm exists between the simple instructions required for crystalline order and the vast number of instructions contained in DNA:

> To describe a crystal, one would need only specify the substance to be used and the way in which the molecules were packed together. A couple of sentences would suffice, followed by the instructions "and keep

on doing the same," since the packing sequence in a crystal is regular . . . .It would be quite impossible to produce a correspondingly simple set of instructions that would enable a chemist to synthesize the DNA of an *E. coli* bacterium. In this case, the sequence matters. Only by specifying the sequence letter-by-letter (about 4,000,000 instructions) could we tell a chemist what to make. Our instructions would occupy not a few short sentences, but a large book instead![115]

Hubert Yockey also rejects the crystal imperfection hypothesis on the basis of information theory:

The transfer of information from clay surfaces to organic macromolecules that is presumed to be a pseudo-DNA/RNA/protein system is mathematically *impossible*, not just unlikely, if the entropies of the two probability spaces are not equal. To say that crystal life is a modified perfection while molecular life is a tamed chaos is merely a play on words. . . . The clay scenario is one of the attempts to use the "order" that is characteristic of a crystal as an analogue of the "order" that is supposed to characterize informational biomolecules . . . The progression of the sequences derived from clay to proteins is essentially the same process conceived in the origin of life by chance. . . . Therefore, for this reason also, the clay scenario provides no pathway from the crystal imperfections in clay particles to information biomolecules.[116]

4.1.3.3. Deep sea hydrothermal vents

John B. Corliss of NASA's Goddard Space Flight Center was a member of a team of scientists who discovered several hydrothermal vents supporting life on the bottom of the sea near the Galápagos Islands. Bacteria, tube worms, and clams obtain their source of energy for life not from light but from sulfur compounds flowing from these vents. Dozens of similar vents have been discovered since the Corliss team found the first vent in the late 1970s. Almost all of these vents have been discovered near underwater ridges formed at the seam of two tectonic plates.[117] Corliss argues that vents could have supplied (a) the energy flow required to remove the area near the vents from equilibrium and the effects of the Second Law of Thermodynamics and (b) the nutrients to allow for the creation and sustenance of life. The interior of the vents would act as a womb for life, protecting the first form of life from the meteorites and surface atmosphere which was so hostile to life on the early earth.

Carl R. Woese of the University of Illinois determined that all living organisms belong to one of three branches of life: eukarya (plants and animals), bacteria, and archaea. According to Woese, the first branch evolved from single-celled organisms with a nucleus, the second branch evolved from single-celled organisms with no nucleus, and the third branch is a single-celled organism which up until recently has been confused with bacteria. The confirmation of archaea as a third branch of life was made by Craig Venter of the Institute for Genomic Research, who decoded the genes of a member of the archaea kingdom called *Methanococcus jannaschii*. Archaea live on deep sea vents in near boiling waters, inside volcanos and in bubbling hot springs. They can survive a temperature as high as 120°C. Certain archaea prefer sulfuric environments without oxygen, such as the environment at a deep sea hydrothermal vent.

Stanley L. Miller and his colleague Jeffrey L. Bada are not convinced that deep sea hydrothermal vents could have served as the womb for the origin of life. They have conducted experiments which indicate that the incredibly hot water inside these vents (frequently exceeding 300°C) would destroy complex organic compounds. Miller has said that if the surface of the early earth was a frying pan, then the deep sea hydrothermal vent was a fire.[118] Norman R. Pace of Indiana University does not believe that the first organisms could have originated at these vents. He disagrees with James Corliss and hypothesizes that the archaea originated in another place, perhaps near the surface of the earth during a respite from the meteorite impacts, and then migrated to the vents.[119]

Günter Wächtershäuser, a patent attorney practicing in Munich with a doctorate in organic chemistry, speculates that life started as a metabolic process on the surface of pyrite or ferrous sulfide. Pyrite is comprised of one iron and two sulfur molecules. Such a surface is positively charged, and the continuing formation of pyrite from iron and sulfur produces energy as electrons which encourage organic compounds to react with one another and polymerize. He has hypothesized that the first cell could have been a grain of pyrite enclosed in a membrane. According to this theory, the cell could have replicated if the pyrite grain produced a crystalline "bud" which in turn became enclosed in its own membrane and broke free from the first cell.[120]

Recently Wächtershäuser and his colleague Claudia Huber of the Technical University of Munich conducted an experiment demonstrating that, with the assistance of ferrous and nickel sulfides, gases known to exist in deep sea hydrothermal vents provide a method of chaining carbon atoms together and of forming acetic acid and an activated form of that acid called thioesters.[121] Thioesters play a role in cellular metabolism. Christian De Duve hypothesizes that thioesters could have formed protoenzymes which eventually could have become RNA which may have acted as the predecessor of DNA. This proposal, however, only brings us back to the problems of the RNA world—the concept that RNA acted as a catalyst for the origin of life which appears to be a dead end based on the improbabilities discussed in section 4.1.3.4.

Moreover, Wächtershäuser and De Duve's speculations do not assist in solving the mystery of the generation of information in inorganic molecules. The demonstration of a natural method of chaining carbon atoms is interesting and demonstrates how a certain order could occur, but it ignores the question of complexity as defined as the generation of sufficient information content necessary to replicate and maintain the structure. Again, complexity and not order is the issue in origin of life scenarios. The experiment generates a multi-dimensional mono-molecular layer which is not sufficient for the generation of information content, because all informational molecules are one-dimensional. The experiment also does not address the sequence issues raised in this book. An appeal to the RNA world is a failed paradigm for the reasons stated above so a plausible pathway to the genetic code is not adequately addressed.[122] Finally, the speculations assume that the answer to life's origin lies in the laws of physics and chemistry. In addressing Harold Morowitz's conjectures in the next section, we discuss why the origin of the information content of the genetic code is not likely to be found ultimately in these laws.

4.1.3.4. Metabolism recapitulating biogenesis

Harold Morowitz, having performed the probability calculations described above, rejects the chance origin of life scenario and proposes a scenario based on his belief in the self-ordering power of the elements in the periodic table. Morowitz does not consider the primeval soup to be a plausible paradigm and speculates that life is a natural extension of the laws of physics and chemistry. He postulates that intermediary metabolism of autotrophs recapitulates prebiotic chemistry and describes a Universal Metabolic Chart which at its core could once perform fundamental reactions without enzymes. Rather than beginning with RNA, DNA, or protein synthesis, he hypothesizes that the transition from non-life to life begins with the closure of an amphiphilic[123] bilayer membrane into a vesicle containing water soluble and water insoluble components. He rejects Monod's contention that chance events imply that the origin of life is unpredictable, but rather speculates that the origin of life is a deterministic result of the laws of chemistry.

Morowitz emphasizes the principle of continuity in his theory which requires that the chemistry and biology of contemporary living cells should contain at least vestiges of a proposed origin of life scenario before such a scenario can be taken seriously. Consequently, he dismisses the hypothesis of clay or pyrite related scenarios:

> The principle of continuity may be introduced to critique proposed theories asserting that microstructures of clays or other minerals such as pyrite were essential elements in the transition to life. Since no clay structures or vestiges of clay structures exist in contemporary cells and since nothing in the logic of

clay chemistry is unique, the introduction of the clay
hypothesis violates continuity without persuasive arguments for the logical necessity for such a violation.
The introduction of clays or pyrites, as other than
boundary structures, needlessly complicates origins
of life theory. [124]

Morowitz considers the formation of closed vesicles to be a
major event in the origin of life. Under his theory the emergence
of a lipid enclosed vesicle created a distinct environment in which
metabolism could form. He speculates that these vesicles form
when an amphiphilic molecule combines in water with another
amphiphilic molecule with their hydrophoic ends joining in pairs.
These pairs then combine to become sheets which form a vesicle
with a primitive membrane which establishes a physical separation between the components of the vesicle and the outside environment. The amphiphilic molecules of the membrane would
have their polar ends facing out into the aqueous environment
and their non-polar ends facing the interior of the vesicle. The curvature of the membrance causes an influx and efflux of particles
through the membrane. Gradually the interior of the vesicle and
the outside environment have different compositions.

He postulates that the earliest prebiotic vesicles were photo-
synthetic. Among the molecules dissolved in the interior of the
vesicle are chromophores, molecules capable of absorbing light.
Using the sun's energy, the chromophores transform the vesicles
into energy transduction devices converting light into electrical
potential energy. These vesicles, with an electrical potential
maintained by the energy of light, would operate without the
benefit of amino acid catalysts, but pursuant to thermodynami-
cally driven reactions favored by the periodic table of elements.
Phosphorylated compounds would be possible in these vesicles
which can generate coupled keto acids. In the presence of ammonia, these acids could ammonify and convert into amino acids.
He assumes that the entry of ammonia into the intermediary metabolism in the vesicle results in amino acids and small peptides
which adsorb on the membrane's surface and perform catalytic
functions. Morowitz suggests that the order of cellular formation
may be consistent with the following:

The pathway of nitrogen influx in biosynthesis almost universally starts with the reaction of ∞-ketoglutarate with ammonia to form glutamic acid. A second reaction generates glutamine, and all other nitrogen compounds are generated by group transfer reactions from these starting materials. Purines, pyrimidines, and nucleotide coenzyme systems are then synthesized using various amino acid precursors, suggesting the order of evolution to be metabolic intermediates, amino acids, purines and pyrimidines, cofactors, RNA, and DNA. It is in this sense that I assert that metabolism recapitulates biogenesis. In somewhat more detail,

> lipids (vesicles) → phosphorylated compounds
> → keto acids → amino acids → nitrogen bases
> → cofactors → coding molecules[125]

His conjecture that in autotrophic prokaryotes and photo-synthesizers intermediary metabolism recapitulates prebiotic chemistry is a novel approach to the origin of life. Morowitz admits that the stage of the genetic code is the most difficult to understand. He postulates that the earliest coding system involved only RNA and that DNA, transcription and other aspects of the code appeared later. His theory differs from many others in that the code is postulated to be a later event in the origin of life process rather than an early event.[126] Thus in his model cells originate first, proteins, RNA and the genetic code follow. George Johnson, in his recent book, *Fire in the Mind*, summarizes Morowitz's theory as follows:

> Viewed panoramically, Morowitz's origin myth has a compelling logic to it. Life, in his view, arose through a series of levels, each more complex than the last. First were empty vesicles dividing and fusing like oil drops, then vesicles with simple chemistries inside. Among these were vesicles with the means for making their own components. When one of these cells "discovered" nitrogen, the next step was enzymes and the richer chemistries they entail. Finally came

the enzymatic production of nucleic acids. With this development, the cells had the ability to keep a separate record of their genetic information; they could mutate and evolve. If Morowitz is right, the potentially unending regression . . . bottoms out in the laws of chemistry, which arise, in turn, from quantum mechanics. In the end, it is simple physics that gives rise to . . . the vesicles. Providing a buffer against the randomness of the environment, they allow for the formation of the delicate chemical arrangements which otherwise would be unlikely to emerge at all.[127]

Morowitz emphasizes that the study of life's origins is a laboratory science. Unlike his Santa Fe Institute colleague, Stuart Kauffman, who mainly works with abstract computer simulations, Morowitz works with laboratory experiments. He has demonstrated that the pathway for cells to assimilate nitrogen can function without protein catalysts. In one experiment he and Sherwood Chang showed that the entry of nitrogen into intermediary metabolism (NH 3 + $\propto$-ketoglutaric acid $\rightarrow$ glutamic acid) can be carried out in the absence of enzymes. Much more evidence, however, is required to give credence to his hypothesis that the fundamental laws of chemistry and physics enable life to function without enzymes.

Although Morowitz has structured a novel approach to the origin of life, he has not solved the difficulties raised by information theory. The obstacle of the method of generating sufficient information remains. *The paradigm for the emergence of life contains algorithms which must have at least as much information content as the genetic messages they claim to generate. The method for such generation is not clear. Because the information content or complexity in the laws of physics is much less than the content in the genome, the gap in content must be explained.* The information generation is not likely to flow from the laws of chemistry and physics alone. As Yockey has stressed:

> The reason that there are principles of biology that cannot be deduced from the laws of physics and chemistry lies not in some esoteric philosophy but simply in the mathematical fact that the genetic infor-

mation content of the genome for constructing even
the simplest organisms is much larger than the information content of these laws. Chaitin has examined
the complexity of the laws of physics by actually pro-
gramming them. He finds the complexity amazingly
small.[128]

Morowitz's search for self-ordering power in the elements of
the periodic table encounters many obstacles. His conjecture has
a large chasm to cross in exploring the genetic code. The laws of
chemistry do not appear to contain the answer. The remarkable
capacity for the DNA molecule's transmission and storage of information is possible because of DNA's flexibility for constituting
the symbols of DNA in a vast multitude of sequences. The pattern of DNA is not regular and predictable, but extremely flexible
so that the base sequence of a DNA molecule is similar to sentences formed from the Roman alphabet. Just as the opening sentence in this book does not determine all of the content that
follows, so the beginning base in DNA does not determine the
pattern of the following bases. In this book rules of grammar put
some required form around the writing but do not control its
content. Similarly, flexibility and lack of a regular, predictable
pattern in DNA argue against an inherent law in the chemistry of
the elements of DNA. In the words of Nancy Pearcey and Charles
Thaxton:

> A law produces regular, predictable patterns. Recall
> our earlier discussion of proteins. Biologists originally
> hoped to find a general law of assembly for proteins.
> And how did they expect to discern the effects of a
> law? They looked for regularities, patterns. It was
> when geneticists failed to find an overall pattern that
> they realized that they were dealing with something
> not produced by natural law.
>
> The same reasoning applies to DNA. If we were to
> find regular, repeating patterns, that would consti-
> tute evidence of an underlying law. But a repeating

pattern encodes little information. Computer buffs sometimes like to create wrapping paper by commanding the computer to print "Happy Birthday!" again and again until the page is filled. The result is a repeating pattern that conveys very little information; the entire page conveys no more information than the first two words.

If the origin of the DNA sequence were a material force, such as chemical bonding forces, then we would get something analogous to computer-generated wrapping paper. The entire DNA molecule would consist of repeating patterns, which would encode very little information.[129]

Michael Polanyi, the former Berkeley and Oxford professor who held doctorates in physical chemistry and medicine, noted that the genetic code would be impossible if the order in the items in the DNA molecule were chemically necessary. If DNA were so highly ordered, it would not be able to carry more than one instruction and would not be able to transmit the vast number of instructions employed by the genetic code. A DNA molecule with a high order due to a strongly bound chemical structure would only be able to enter into the type of relationships which exist among all ordinary molecules and would not be able to enter into any communicative (linguistic) relationship with other molecules. In his article entitled, "Life Transcending Physics and Chemistry," Polanyi explained his position:

All chemical compounds consist of atoms linked in an orderly manner by the energy of chemical bonds. But the links of a compound forming a code are peculiar. A code is a linear series of items which are composed, in the case of a chemical code, of groups of atoms forming a chemical substituent. In the case of DNA, each item of the series consists of one out of four alternative substituents. In an ideally functioning chemical code—to which I shall limit myself—each alternative substituent forming a possible item of the series must have the same mathematical chance of appearing at any point of the series. Any difference of alternative chances would reduce the amount of information transmitted, and if there were a chemical law which determined that the constituents can be aligned only in one particular arrangement, this arrangement could transmit no information. Thus in an ideal code, all alternative sequences being equally probable, its sequence is unaffected by chemical laws, and is an arithmetical or geometrical design, not explicable in chemical terms.[130]

Chemical structures formed by the stabilizing effects of chemical bonds cannot have any significant amount of information content. DNA can function as a code only if its base sequence is not determined by physical and chemical laws. Polanyi maintained that "all objects conveying information are irreducible to the terms of physics and chemistry."[131] He made an analogy between DNA transmitting information and a book transmitting information. Just as the operation of the book is not reducible to chemical terms so the operation of the DNA molecule cannot be described by chemical laws. In his words: "As the arrangement of a printed page is extraneous to the chemistry of the printed page, so is the base sequence in a DNA molecule extraneous to the chemical forces at work in the DNA molecule."[132]

The base of sugars and phosphates in the DNA molecules are chemicals, but the sequence of these bases are not explained by the laws of chemistry or physics. The flexibility in these bases is similar to the flexibility in the alphabet letters comprising a poem. A poem is distinct from a random organization of alphabetical symbols because of the message and art conveyed. The alphabet letters selected to produce a poem are not determined by the chemicals in the pencil, pen or computer print cartridge used to write the poem, nor are the messages of the sequences of bases in the DNA determined by information inherent in the chemical elements which constitute these bases.

The poem and its information content is independent of the type of substance used to write the poem. The poem could be written in chalk, ink, paint, ice or any of a wide variety of materials, but the message of the poem is not dependent upon the materials comprising the writing. Similarly, the information in the DNA molecule is independent of the bases of sugars and phosphates which comprise the molecule. If information is independent from these chemicals, the information did not arise from the chemicals; just as a poem written on a blackboard did not arise from the chalk.

The answer for the origin of the information content of the genetic code is not likely to be found ultimately in the laws of chemistry. Pearcey and Thaxton present the argument:

> The sequence of bases that spell out a message in the DNA molecule is chemically arbitrary. There is nothing intrinsic in the chemistry of any base sequence that makes it carry a particular meaning. In fact, there are many base sequences possible besides the ones actually used in the cell—all of them equally probable in terms of chemical forces. By merely examining the physical structure, you could not detect any difference between these useless base sequences and those necessary for life. There is nothing in their physical make up that distinguishes the two sets of molecules. Out of a vast number of possible base sequences, somehow only a few carry meaning. . . . If the physical components of the DNA molecule are not distinguished in any way, then it seems clear that no analysis of the physical components can explain what makes it unique—what makes it function as a symbol system. Instead, the answer is found in the analogy between DNA and a written message. What confers meaning on particular sequences of letters in a message are linguistic conventions—rules of usage, grammar, and sentence structure.[133]

4.1.3.5. Complexity on the edge of chaos

Stuart Kauffman is attempting to devise a theory in which the principles of self-organization are imposed from within an organ-

ism's internal laws, rather than from outside the organism. Kauffman believes that complex systems arise on the "edge of chaos"
where forces of order exist. John Horgan defines this phrase:

> The basic idea is that nothing novel can emerge from
> systems with high degrees of order and stability, such
> as crystals. On the other hand, completely chaotic
> systems, such as turbulent fluids or heated gases, are
> *too* formless; truly complex things—amoebae, bond
> traders and the like—appear at the border between
> rigid order and randomness.[134]

Unfortunately, the definition of the term "complexity" is a
moving target with Kauffman and his Santa Fe colleagues continually changing the definition. Seth Lloyd of the Massachusetts
Institute of Technology and the Santa Fe Institute compiled a list
of at least 31 definitions of complexity used by persons connected
with the Institute and with complexity studies. Such a variety of
meanings makes any analysis of the "complexology" of Kauffman
and his Santa Fe colleagues difficult. As Horgan notes:

> At various times, researchers have debated whether
> complexity has become so meaningless that it should
> be abandoned, but they invariably conclude that the
> term has too much public relations value. Complexol-
> ogists often employ "interesting" as a synonym for
> "complex". But what government agency would sup-
> ply funds for research on a "unified theory of inter-
> esting things"?[135]

Kauffman bases his ideas of complexity on computer simulations which, compared to laboratory experiments, make his theories a "fact free" science. He proposes that complexity initiated
self-organization is a principle of nature, so that life emerged
from a mixture of various molecules in the prebiotic soup which
reached a certain level of complexity and then self-organized
pursuant to laws. In certain respects he joins Cairns-Smith and
others in confusing complexity as a derivative of order in a system far from equilibrium. Many of his statements concerning

complexity existing on the edge of chaos may only constitute a play on words. His propositions remain speculative computer simulations without proven answers to the questions raised by information theory and by Polanyi and others on the irreducible aspects of information. Stanley Miller and other scientists are critical of Kauffman's lack of laboratory experiments. In their view computer simulations cannot take the place of laboratory work. Computer simulations may create a framework for thought, but, as Morowitz emphasizes, cannot serve as a substitute for experiment and observation. Kauffman uses a mathematical analysis which reduces the special characteristics of organisms to mathematical symbols as he manipulates the symbols.[136] He and some other Santa Fe complexologists base their positions on the following syllogism:

> There are simple sets of mathematical rules that when followed by a computer give rise to extremely complicated patterns. The world also contains many extremely complicated patterns. Conclusion: Simple rules underlie many extremely complicated phenomena in the world. With the help of powerful computers, scientists can root these rules out.[137]

This reasoning was discredited by Naomi Oreskes of Dartmouth College and her colleagues in an article written in the February 4, 1994, edition of *Science*. Oreskes argued that the verification and validation of numerical models of natural systems was impossible, because natural systems are never closed. Oreskes argued that it is impossible to demonstrate the truth of any proposition except in a closed system based on pure mathematics and logic. Her argument may be summarized in part by the following example:

> "If it rains tomorrow, I will stay home and revise this paper." The next day it rains, but you find that I am not home. Your verification has failed. You conclude that my original statement was false. But in fact, it was my intention to stay home and work on my paper. The formulation was a true statement of my intent. Later,

> you find that I left the house because my mother died,
> and you realize that my original formulation was not
> false, but incomplete. It did not allow for the possibility
> of extenuating circumstances. Your attempt at verifica-
> tion failed because the system was not closed.[138]

Despite Oreskes's and other challenges, Kauffman concludes from his computer simulations that when a system of simple chemicals reaches a particular level of interconnectedness, the system experiences a transition or dramatic phase change whereby molecules spontaneously combine into larger and more complex molecules with catalytic capability. Kauffman calls this process "autocatalysis" and argues that it leads to life.[139]

Veritas temporis filia.[140] Time will tell if his search for self-organizing laws will be able to give a successful explanation for a method for the generation of sophisticated information content into inert matter. To begin with a computer which already has a great quantity of information content and then perform "random" simulations begs the question. A large quantity of complexity or information exists in the machine which produces the simulation.[141] This is not the environment in the real world of the periodic elements. The *vexata quaestio* remains the formation of the genetic code. At present and for the purposes of the questions presented, it is sufficient to note Kauffman's efforts and that he recognizes the difficulty with the proposition of life emerging from accident alone. His conjectures encounter the same problems concerning the principles of biology and the laws of physics and chemistry noted above, including Yockey's argument that the laws of biology cannot be deduced from the laws of physics and chemistry, because the information content in the laws of physics is much less than the information content in the genome.[142]

4.1.4. ALH84001

Allan Hills 84001 (sometimes below referred to as ALH) is a 4.2 pound piece of rock found in 1984 in a field of jagged ice in the Allan Hills region in South Victoria Land of Antarctic by a National Science Foundation meteorite team. The rock is about 4.5 billion years old and probably originated on Mars. About 16 million years ago a large asteroid hit Mars and sent ALH into space. The rock fell onto the ice near the South Pole between 13,000 and 15,000 years ago. At a press conference on August 7, 1996, NASA introduced a team of scientists who contend that they may have found evidence of life in the rock which appears to have come from Mars. This announcement was exactly 20 years from the date NASA announced that it had detected signs of "activity" in samples of Martian soil. With respect to the earlier announcement, further examination demonstrated a total absence of evidence of life in those samples. Will ALH84001 prove to be different? The jury may be out for several years.

In the August 16, 1996 edition of *Science*, David McKay published the now controversial report. McKay encouraged caution:

> We are not claiming that we have found life on Mars, and we're not claiming that we have found the smoking gun, the absolute proof, of past life on Mars. We're just saying we have found a lot of pointers in that direction.[143]

William Schopf, a bacteria expert at the University of California at Los Angeles, is very skeptical of the biological conclusion. "At this point, in my opinion, the biological interpretation is probably unlikely," Schopf commented at the August 7, 1996, news conference. Derek Sears, editor of the journal *Meterorites and Planetary Science*, also concluded: "there are nonbiological interpretations of McKay's data that are much more likely."[144] Much work needs to be done to determine exactly what is contained in ALH.

The argument for and against evidence of life centers around four findings.[145] Each finding has alternative explanations so the debate will be interesting. The evidence as of early March, 1997, may be summarized as follows:

Carbonate Globules. ALH contains carbonate globules in its cracks. These globules are similar to carbonates associated with ancient bacteria on earth. On the other hand, these globules and the minerals in them could have formed through known inorganic processes. There are processes whereby inert matter forms carbonate globules. The formation of these globules with the assistance of bacteria is impossible unless the globules formed at low temperatures. As Christopher F. Chyba notes: "If the carbonates in ALH84001 were formed at high temperature in an impact event, a biological interpretation would fail."[146] The probability of a biological interpretation was diminished by the recent findings of Harry McSween, Jr., Ralph Harvey and John Bradley, one of the premier analysts of microscopic material in geology. John Wilford summarized their findings on the temperatures at which globules formed: "In their examination of the supposed fossils, the scientists said they found that surrounding minerals probably formed from vapors that crystallized at temperatures as high as 1,400 degrees Fahrenheit, conditions much too hot to have included biological processes."[147] (See discussion below on magnetite.)

At present, we are left with conflicting values for the temperature for formation of these carbonate globules. Major element chemistry appears to require a formation temperature exceeding 500°C. Oxygen isotope composition, however, indicates a temperature less than 100°C. Until these differing interpretations are resolved, a low temperature consistent with a biological origin for the carbonate globules is not certain.

The date when the carbonates formed is also subject to differing interpretations. One report implies a date as late as 1.4 billion years ago, while McKay's team set a date of 3.56 billion years. The date is important in determining the atmosphere on Mars, including the question of the presence of water, at the time of the carbonate formation.[148] Kenneth Nealson of the University of Wisconsin at Milwaukee, an authority on bacterial carbonate precipitation, whose work was cited by McKay and his team, warns that warm fluids circulating through the crust of Mars could have deposited the same sequence of minerals without the involvement of any organism.

Jim Papike and Charles Shearer of the University of Mexico examined the iron disulfide in the fractures of ALH and could not find any ratio of sulfur isotopes which would be consistent with known biological activity.[149] Although pyrite's presence in ALH can be explained by the hypothesis of Martian bacteria, that hypothesis is extremely improbable. Although bacteria on earth make pyrite from the digestion of sulfates, bacteria works mainly on light sulfates with sulfur atoms known as ^{32}S, rather than the heavier sulfur with stronger chemical bonds known as ^{34}S. Earthly sulfur is approximately one third ^{32}S and two-thirds ^{34}S, but ALH has a higher ratio (by 50 atoms per 1,000) of ^{34}S than the ratio found in earthly bacteria or remnants of bacteria. In other words, the heavy ^{34}S rather than the light ^{32}S predominate in ALH with an excess of five to eight atoms of ^{34}S per 1,000 atoms compared with bacteria from earth. This ratio is inconsistent with biological activity. One comment on this finding in an article entitled, "Fool's Gold on Mars?," concluded:

> This measurement, completed in December, 1995, was not intended as a riposte to last month's life-on-Mars announcement. Its aim was to explore Martian geology and test a new method of weighing atoms in rock samples. *Still, its authors conclude that the pyrite is not from bacteria.* The extra ^{34}S could, they think, have come from weathering processes when the meterorite was part of the surface of Mars, or because of something unusual about the crystallization of pyrite in this case.[150]

One might argue that the composition of Martian sulfur differs from that of earth's sulfur. Again, the hypothesis is very improbable; most of the solar system has the same inventory of sulfur ratios. Other meteorites, one of which also came from Mars, contain the same ratio of ^{32}S to ^{34}S as found on earth. In fact, sulfur samples from meteorites are so uniform that the world standard for sulfur ratios is based on the composition of a meteorite that landed near Flagstaff, Arizona. Monica Grady and her colleagues find too much evidence to remove their skepticism:

> . . . the carbon isotope composition of the material isn't sufficiently different from terrestrial organic compounds to provide an *unambiguous* signature for the planet of origin of the molecules.[151]

Polycyclic Aromatic Hydrocarbons. The carbonates in ALH occur in areas rich in polycyclic aromatic hydrocarbons (PAHs).

These PAHs are not good biomarkers or signs of life in ALH because they are not directly synthesized in biological systems but produced by a process of metamorphism. ALH is not metamorphosed.

PAHs can be formed when organic matter decomposes. Coal, for example, is made of the fossils of plant life. But PAHs are commonplace in interplanetary and interstellar dust particles and in meteorites from the asteroid belt. These PAHs are the residue of non-biological reactions among carbon compounds. Even on earth, PAHs are ubiquitous in their presence and are formed not only by the decomposition of living matter, but also by power plants and automobile engines. PAHs are present in practically every gas cloud in the Milky Way galaxy. John Kerridge of the University of California, Los Angeles, comments that there are plenty of explanations for the PAHs that don't require life: "Decompositions could certainly produce polycyclic aromatic hydrocarbons, but there are dozens of other mechanisms for making PAHs."[152] He notes that PAHs could have formed from simpler compounds on Mars which were merely chemical and inorganic. Kerridge describes an applicable hydrothermal process: "Imagine hot fluids flowing through the crust. The crystallization of magnetite, iron sulfides and carbonate with a change in chemistry over time is perfectly reasonable. If anywhere in the subsurface of Mars, there are PAHs, then they would be carried by this fluid and deposited where the fluids crystallize. I think the nanostructures are most likely an unusual surface texture resulting from the way in which the carbonate crystallized."[153] Bernd Simoneit, a chemist at Oregon State University, agrees: "Hydrothermal synthesis could take inorganic carbon and water and make aromatic organics; you would get the same ones they report."[154] One must also remember that Mars and earth were constantly exchanging matter through meteorites and the solar wind for millions of years. Meteorites and the solar wind could have transferred PAHs to the surface of Mars and to ALH and many other rocks.

Magnetite. The rims of the carbonate globules have tiny crystals of magnetite, made from iron and oxygen in a shape and composition similar to magneto-fossils made by bacteria on earth. On the other hand, magnetite is a common mineral which can be formed under a variety of circumstances, including inorganic precipitation. John P. Bradley and his colleagues reviewed the "nanofossils" with an electron microscope magnifying the structures 500,000 times their actual size. Bradley noted that the magnetite was not in the formation normally seen in bacteria where the magnetite residues appear as a chain of crystals similar to a string of pearls. In ALH the crystals were elongated and cigar shaped and contained a spiral defect which is a known product when crystals are formed at temperatures too high to be consistent with biological processes.[155] Bradley and his colleagues noted that the magnetite grains in the meteorite had unique forms and structures with whiskers and platelets and defect structures with screw dislocations which were consistent with an inorganic formation "from a hot vapor or supercritical fluid but inconsistent with an origin from biogenic precipitates or microfossils."[156] They concluded that the screw dislocations and magnetite whiskers were clearly inconsistent with biogenic magnetite. The whiskers and platelets were not signs of life. As Bradley and his colleagues wrote:

> Based on natural and synthetic occurrences, it is likely that at least some and possibly all of the magnetite in ALH84001 formed at temperatures of 500-800°C from a vapor or supercritical fluid. Such a mode of formation is consistent with inorganic precipitation from a volcanic- or impact-derived gas or fluid, but not with a biogenic origin (the maximum temperature for terrestrial biota is ~ 120°C)[157].

On the other hand, two independent studies reported in the March 14, 1997, issue of *Science* indicated that the formation of the carbonates was at a temperature between 20 and 80°C. Joseph Kirschvink and his colleagues noted that two adjacent pyroxene grains in the meteorite possessed a stable natural remanent magnetization, implying a magnetic field on Mars when the grain cooled. Because the natural remanent magnetization directions from the particles differ, these researchers concluded that ALH84001 was not hot when the carbonate globules formed and that the globules probably formed at low temperatures.[158] Similarly, John W. Valley and his team concluded that the isotopic variations and mineral compositions did not provide evidence of a temperature higher than 650°C and actually suggested non-equilibrium processes at a temperature less than 300°C.[159]

Ralph Harvey of Case Western Reserve University, however, said that, despite any other studies, his work with John Bradley in finding whiskerlike defects posed such a serious problem that it was tantamount to conclusive proof of a high temperature formation for the magnetite minerals in the carbonate and precluded a biological origin.[160] Laurie Leshin of UCLA appears to agree with Harvey and Bradley because of the result of a study her team conducted with an ion microprobe which suggested that either there was not enough water to support life when the carbonates formed, or the water was too boiling hot, or the fluid changed with the formation of the carbonates. The latter option was not conducive to life, but leaves the matter not completely resolved.[161]

To continue the pros and cons in the debate, we should note that Harold Morowitz is skeptical and considers the magnetite to be weak evidence for an organic explanation. He evaluates the assumption that single-domain magnetite found on the meteorite is biological in origin, given differences between the magnetic fields of earth and Mars:

> Particles on magnetite are characteristic of some bacteria on Earth (such as *Aquaspirillum magnetotacticum*), where the strong magnetic field invites a navigational role for permanent magnets. An ecological role for magnetite in the weak magnetic field of Mars is less persuasive.[162]

Tube Shapes.   Everyone at present acknowledges that the tube shape configuration is the weakest of the four pieces of evidence. Tube shaped objects are on the carbonate globules. These objects are smaller than one hundredth of a diameter of a human hair. One of them looks segmented, divided in the appearance of an earthworm. The non-segmented objects have an appearance similar to tiny bacteria. On the other hand, these images could merely be dried clay or odd shaped crystals. Many inorganic processes could produce these shapes. At 20 to 100 nanometers in length the tube shaped objects are one-one hundredth of the diameter of a human hair and most are approximately one-one thousandth of such a diameter. I am writing on March 20, 1997, and until a discovery in the last few months, these shapes were 100 times smaller than the smallest microfossils of bacteria ever found. Kathie Thomas-Keptra, however, recently reported that samples from drillings in the basalts of the Columbia River in Washington showed similar size microfossils with similar shapes. To appreciate the incredibly small size of these shapes, one must understand that one thousand of them lying end to end would equal the diameter of the period following this sentence.

Kenneth Nealson is skeptical: "The little blobs (in ALH84001) didn't convince me. I think you can form little blobs on rocks with all kinds of chemical precipitates."[163] Jack Farmer, a NASA exobiologist, adds: "The problem is that at that scale of just tens of nanometers, minerals can grow into shapes that are virtually impossible to distinguish from nanofossils.[164] Harold Morowitz calculates that the tube-shaped objects can only contain approximately one hundred million atoms. A hundred million atoms, however, arranged in an optimum structure is not a sufficient number of atoms for a simple bacterium.[165] (His view may not be the same, however, after he examines the Columbia River microfossils.) Commenting in the September 20, 1996 issue of *Science*, Morowitz writes:

> To suggest that objects on the order of 10^{-22} stere (10^{-22} cubic meters) could be cellular requires that all of the necessary biological functions of a cell could be carried out by an "organism" with less than 100 million atoms. Such an "organism" would be two orders of magnitude smaller than the smallest known one-celled organisms on earth, mycoplasma.[166]

In summary, the carbonate globules, magnetite, iron sulfide and PAHs are all generated in large quantities by inorganic and organic processes. The debate whether ALH contains real evidence of life will probably proceed for a long time. It is important to remember that, as far as we know at present, ALH does not contain any fossils of living matter, but only some compounds which are associated with fossils of living matter. Skeptical scientists will want to see some evidence of a cell wall. As Harold Morowitz asserts, "The only life we know for certain is cellular."[167] Monica Grady and her colleagues note:

> . . . one of the most basic criteria applied when identifying microfossils is the presence of organized elements, such as a cell wall. The structures shown . . . do not appear to demonstrate any such features. The leap from molecules to microfossils is an enormous one in terms of biological evolution, and there is probably not enough evidence yet for that leap having been made on Mars.[168]

The most decisive test for ALH84001 may be the search for a cell wall. Scientists are preparing for this test which may prove to be the most meaningful examination of the meteorite. The problem at present is that the instruments currently available may not be entirely up to the task. We may still be many years away from knowing the true significance of ALH84001.

Hugh Ross believes that life or the remains of life will eventually be discovered on Mars, because Mars is only thirty-five million miles away from earth. There are many reasons to believe that millions of earth's microorganisms have been deposited on Mars. These organisms are capable of surviving the conditions in outer space for sufficient time to reach Mars. Meteorites and the solar wind have moved vast amounts of matter from the earth to Mars over the past four billion years. The discovery of life or the remnants of life on a Martian meteorite would not prove the spontaneous generation of life. As Hugh Ross writes:

> Meteorites large enough to make a crater greater than 60 miles across will cause Earth rocks to escape Earth's gravity. Out of 1,000 such rocks ejected, 291 strike Venus, 20 go to Mercury, 17 hit Mars, 14 make it to Jupiter, and 1 goes all the way to Saturn. Traveling the distance with these rocks will be many varieties of Earth life.[169]

For the purposes of this book the question whether ALH actually presents convincing evidence of life on Mars is not essential. It is also not crucial whether this life formed on Mars alone or was first transported to Mars from earth. Extra-terrestrial life may exist,[170] but not by chance alone. The mathematical probabilities discussed in this book argue against life arising by accident on earth or Mars. The ingredient still missing in all scientific origin of life scenarios is an explanation for the generation of information content into inert matter, whether that matter is on Mars or on earth. The absence of any plausible explanation is a gap which would still exist even if giraffes were found on Mars. ALH does not provide such an explanation or give support to the theory that the formation of the universe or the origin of life is governed by chance alone.

The issues discussed in this book remain the same regardless of the results of further investigations on ALH or on Mars. Life appears to be formed only by a guided process with intelligence somehow inserting information or instructions into inert matter. One can ask what purpose would a microorganism have on Mars, but the same question can be asked about a dinosaur on earth. In examining biogenesis theories we must look at the mathematical probabilities, not at metaphysical perspectives, regardless of the way in which they may point. The calculations in this book rule out chance alone for 130 million years or for the entire age of the universe. Something besides chance caused and is causing life.

4.2. Present absence of a plausible scientific theory for generating information content into inert matter

Without evidence for a method of generating sufficient information content in the limited time available, self-organization theories for the formation of life from inert matter are not plausible at the present time. The distinction between living and non-living matter is the existence of a genome or composite of genetic messages which carry the information necessary to replicate and maintain the organism. One may choose on a religious basis to believe in self-organization theories, but such a belief must be based on one's metaphysical assumptions, not on science and mathematical probabilities. Hubert Yockey's conclusion in his 1980 paper, "Self Organization Origin of Life Scenarios and Information Theory," remains valid today:

> Rarely if ever do those who propose an origin of life scenario trouble themselves to make a quantitative estimate of the probability that events in the real world will indeed go as described. . . . The calculations presented in this paper show that the origin of a rather accurate genetic code, not necessarily the modern one, is a *pons asinorum* which must be crossed to pass over the abyss which separates crystallography, high polymer chemistry and physics from biology. The information content of amino acid sequences cannot increase until a genetic code with an adaptor function has appeared. Nothing which even vaguely resembles a code exists in the physico-chemical world. One must conclude that no valid scientific explanation of the origin of life exists *at present*.[171]

PART V

CASE AGAINST ACCIDENT FROM PRECISION OF VALUES IN PARTICLE ASTROPHYSICS REQUIRED FOR THE FORMATION OF LIFE

Having defined life and considered probabilities relating to the formation of life from inert matter by chance processes and discussed the inadequacies of prominent self-organization scenarios, we now turn to an examination of the precision of values and probabilities for the formation of a universe compossible with living matter.

Many proponents of accident or chance as the cause of life proposed their theories when the universe was considered by many scientists to be in a steady state with infinite age. In an infinite, ageless universe, anything can happen. Recent discoveries in modern physics and mathematical analysis accompanying those discoveries, however, changed this view of the universe to that of a young universe, expanding from a definite beginning. From this perspective the probability calculations concerning the precision of many of the values in particle astrophysics present a strong case against an accidental universe. Because we can conceive of a vast number of universes which vary from our own but which would not allow for life, the precision of values in the formation of a universe compossible with life is another way of examining the mathematical probabilities of our universe. The universe appears to be precisely fine tuned for the formation of life. Several lists of "just right" characteristics give significant evidence of design. In referring to some of these lists, Paul Davies wrote:

> Taken together they provide impressive evidence
> that life as we know it depends very sensitively on

the form of the laws of physics, and on some seemingly fortuitous accidents in the actual values that nature has chosen for various particle masses, force strengths, and so on. If we could play God, and select values for these natural quantities at whim by twiddling a set of knobs, we would find that almost all knob settings would render the universe uninhabitable. Some knobs would have to be fine-tuned to enormous precision if life is to flourish in the universe.[172]

5.1. Background foundation for discussion of precision of values in particle astrophysics

To provide a context in which the significance of these precise values can be appreciated, we will first discuss the observations and mathematical analysis which led to the discovery of the expansion of the universe, the Big Bang theory, the COBE satellite evidence for the Big Bang, the singularity at the beginning of the universe and the singularities in black holes, the four fundamental forces of the universe, quantum particle structure, interactions of quantum chromodynamics, proposed grand unified theories, including extra dimensional string theory, and the force and particle activity in the very early universe. With this background we will be able to discuss the many "just right coincidences" which held the formation of life in the balance of the fine details of particle astrophysics. The fine tuning we will explore discloses a universe so remarkably balanced to allow for the origination of life that one may think of such a universe as a finely sharpened pencil standing vertically on its graphite point in a precarious balance. Any deviation in a myriad of physical values would cause the pencil to tilt, fall, and preclude the formation of life.

5.1.1. *Hubble's discovery of the expansion of the universe*

In 1929 Edwin Hubble, a lawyer turned astronomer, working at the Mount Wilson Observatory in Pasadena, California, discovered that the universe was expanding. Using the observatory's 100-inch telescope, Hubble observed that galaxies are expanding away from one another at a velocity directly proportional to their distances apart. (The galaxies close to one another do not actually expand due to gravitational forces between them, but the space between these galactic clusters expands. The universe's expansion is actually the expansion of space itself, a concept that Hubble never really understood.) Hubble observed that the type of light received from spiral nebulae shifted from blue or higher frequencies to red or lower frequencies. The frequency of the wavelength from a source of light will decrease with the speed at which the observer of the light and the source of the light move away from one another. This shift to the lower, red frequencies is understood in the terms of the Doppler effect named after Christian Doppler, a Prague mathematics professor, who discovered that wavelength is affected by motion between the source of light and the observer of the light. When a light source is approaching an observer, the light waves emitted from the source are increasingly compressed. The light wave shifts toward the short wavelength of the spectrum in a blueshift. The opposite is true of a light wave from a source moving away from an observer, and the wavelengths appear longer and produce a redshift. The size of a wavelength shift is proportional to the radial velocity between the observer and the source of the light.[173]

Hubble determined that the Andromeda nebula, the galaxy closest to the Milky Way, was approaching the earth at 50 kilometers per second, but that distant galaxies were receding from the earth at a velocity that increased in a linear proportion to their distance from the earth. Thus, if X galaxy were four times farther from us than Y galaxy, X galaxy would be receding at a velocity four times as great as the velocity of the Y galaxy. Velocity could be calculated by the formula $V = H_o r$, now known as Hubble's law, where r represents the distance and H_o is Hubble's constant, frequently denoted as a ratio of velocity to distance. The precise value of the Hubble constant is still the subject of debate.

The value is important, because it may be used to calculate the age of the universe. To calculate the age of the universe, we simply reverse the process of two galaxies moving away from each other at a certain velocity and determine the time it will take for the galaxies to collide. This computation can be done by the following formula:

$$T_o = \frac{r}{v}$$

where v = velocity, r = the distance between the galaxies and T_o = the time elapsed until the collision of the galaxies. If we replace the velocity with Hubble's law, $v = H_o r$ (where H_o = the Hubble constant) the equation becomes:[174]

$$T_o = 1/H_o = 1/50_{km/s/Mpc} = 20 \text{ billion years}$$

Because the distance between the galaxies is cancelled out, T_o or time is the same for all galaxies and for the universe itself. Because the expansion rate of the universe has actually been decreasing, the time is only an estimate. The equation assumes a uniform rate and would only be precise in a universe devoid of matter and therefore without a gravitational force slowing the expansion.

Wendy Freedman has calculated the value of the Hubble constant at 80 ±/second/megaparsec (a megaparsec is equal to 3.26 million light years) which translates to an age of 8 billion years. This age presents some problems since the oldest objects in our own galaxy, the globular clusters, have ages of about 15 billion years.[175] Gustav Tammann and Allan Sandage measured the Hubble constant at values between 52 kilometers/second/megaparsec and 62 kilometers/second/megaparsec. The average of Sandage and Tammann's values is 57 which translates to an age of 16 1/2 billion years.[176]

5.1.2. The Big Bang theory

Approximately 10 years before Hubble's discovery, Einstein's calculations produced an expanding universe. Because he doubted his equations and was dismayed with the concept of expansion, he added his cosmological fudge factor. Otherwise, the discovery of an expanding universe would have been added to his impressive list of accomplishments. Instead, Edwin Hubble receives most of the credit for the discovery. In addition to Einstein, several other physicists had difficulty accepting the theory of an expanding universe with its implication that the universe began at a finite time in the past from an enormously compressed state. Fred Hoyle ridiculed the theory by calling it the "Big Bang." Proponents and opponents liked the name and its use has dominated the concept of the beginning of an expanding universe ever since.

Despite this name, however, the Big Bang should not be confused with an explosion similar to the detonation of a bomb where pieces of matter are propelled into space. The Big Bang is the expansion of space itself. Pursuant to the theory of general relativity, space is not fixed. The amount of space between galaxies changes over time. The galaxies themselves are not expanding because of the constraints of their gravitational fields, but the space between the galaxies is expanding. The expansion of the universe can be compared to the expanding surface of an inflating balloon. Imagine that small stars or clusters of stars are glued onto the surface of the balloon. As the balloon is filled with air, the surface of the balloon expands and the distance between the stars increases. The balloon's surface does not have an edge or a center. Similarly, the universe does not have an edge or a center.

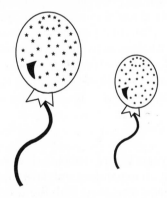

[Figure 1]

Another illustration may be the comparison of the expansion of space to the rising of a loaf of raisin bread in an oven. As the loaf increases in size, the distance between the raisins increases. The space of the universe continually expands in the same manner as a balloon inflates or as a loaf of raisin bread rises. (See Figure 2).

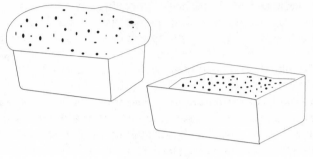

[Figure 2]

5.1.3. Blackbody radiation evidence for the Big Bang

The confirmation of the Big Bang theory was provided by the evidence from the cosmic background radiation. At the Big Bang, the universe had an extremely high density and high temperature. All matter was in thermal equilibrium with radiation. As the universe expanded and cooled to about 3,000° Kelvin, matter and radiation decoupled and expanded separately. The radiation from that time is received by us now with a very large redshift which relates to the enormous distance of many billions of light years. An empirical verification of this radiation occurred in 1964, when Arno Penzias and Robert Wilson, working with an ultra-sensitive radio telescope at Bell Laboratories, detected a background radiation field at a wavelength of 7 centimeters in the microwave region of the radio spectrum where wavelengths are shorter than one meter. This cosmic microwave radiation appeared to come uniformly from the most distant places in the universe. The radiation was an ancient relic from the decoupling of matter and radiation which occurred when the intense heat of the Big Bang dropped to about 3,000°K. As the universe expanded, this radiation was released in all directions and continued to cool to its present temperature of approximately 3°K. In 1949 in their work on the theory of an expanding universe, George Gamow, Ralph Alpher and Robert Herman predicted that the universe's beginning fireball would produce a blackbody radiation of about 5°K. (Blackbody radiation is the hypothetical radiation emitted from a completely black object with the characteristics of the radiation depending only on temperature).

The confirmation of the theory was dramatically made with measurements from the Cosmic Background Explorer (COBE) satellite in January, 1990, when an instrument on board COBE measured the background radiation at 2.726°K which plotted perfectly along a blackbody curve. The measurements deviated from a perfect blackbody radiator by less than one percent from all points in the universe, causing Joseph Silk to comment, "One cannot measure a more precise blackbody in the laboratory than has been detected in the sky."[177]

5.1.4. The singularity ad initium and the singularities of black holes

An expanding universe implies that the universe was previously smaller. If the rate of expansion were reversed, all of the matter in the universe would be compressed to an infinitely dense singular point smaller than a proton. The Big Bang emerged from such a singularity where spacetime is subject to an infinite curvature and does not exist in any terms which can be described by the known laws of physics. Past, future, and present are meaningless terms in this singularity. There is no "before" in this singularity, because time does not exist. Only after the Big Bang at Planck time (10^{-43} of the first second) do space and time exist as we understand those terms. From the Big Bang to Planck time (T_o to T_p) the known laws of physics are inapplicable and no quantum particles exist.

To appreciate the characteristics of the singularity in the Big Bang theory, an understanding of the singularities in black holes provides a useful comparison. To understand a black hole's singularity, we need to describe how high mass stars can collapse to a black hole with a single point of infinite density. When the mass of the burned-out core of a dying star is greater than 3 solar masses ($M_\odot = 3 \times 1.989 \times 10^{30}$kg), the weight of the core compacts matter to densities exceeding nuclear density and even to a density which is infinite where space and time are infinitely distorted. (See discussion below on Einstein's theory of general relativity for an understanding of the curvature and distortion of space and time). Space is so severely curved that nothing, not even light, can escape. The place at which the escape velocity from this distortion of space equals the speed of light is the event horizon. This event horizon surrounds the singularity, a point of infinite density at the center of the black hole. If one imagines a circle with a singularity at its center, the event horizon will be the circumference of the circle, and the radius of the circle will be the distance from the singularity to the location where the escape velocity is equal to the speed of light. A black hole's singularity, its distortion or curvature of space, and its event horizon is illustrated in Figure 3.

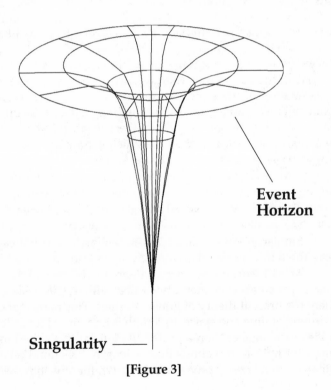

Event Horizon

Singularity ——

[Figure 3]

The length of the radius to the event horizon depends on the mass of the black hole. A nonrotating black hole's distance to its event horizon is known as the Schwarzchild radius, named after the German astronomer, Karl Schwarzchild.[178] As the density of the mass of a black hole increases, the volume increases proportionately to the cube of the Schwarzchild radius. Accordingly, a large black hole has less dense matter than a small black hole.[179] Because light cannot escape the extreme curvature of space or gravitational force within the Schwarzchild radius, events within the radius of the black hole are hidden from outside observation.

To understand the comparison between the singularity of a black hole and the singularity of the Big Bang theory, it may be useful to note the work of physicist Arthur Compton on the wave-particle duality in nature. Light can behave like a particle and like a wave. In dispersion of light by a prism, light behaves like a wave. In the photoelectric effect, light behaves like a particle. Electrons may behave like waves. This apparent contradiction was verified by physicist Compton in his x-ray photons experiments. Electrons act like particles on a very small distance scale (one ten-billionth of a centimeter or 0.243nm).[180]

The laws of quantum physics and general relativity break down in a black hole with a mass so small that its Schwarzchild radius is less than the Compton Wavelength which is the threshold length below which the quantum mechanics of a particle become relevant in relativistic quantum theory. (A proton, for example, has a Compton Wavelength of 2×10^{-14} centimeters.) Imagine the reversal of the Big Bang expansion of the universe and the flow of matter back to a state where it is so compressed that its Compton Wavelength is equal to its Schwarzchild radius. This is the smallest mass where general relativity does not break down. Similarly, the smallest black hole which does not cause a breakdown in general relativity has a mass known as the Planck mass.[181] This Planck mass is equal to about 10^{-5} gram and is also equal to the smallest elementary particle with the highest density where the present theory of gravity applies. This mass marks the beginning of time and space. Its length is known as Planck length and is much smaller than a proton.[182] Planck time is the time required for light to travel this Planck length.[183] General relativity and all known laws of physics break down for any "time" earlier than Planck time. Actually, in the Big Bang theory Planck time is the instant at which the cosmic clock begins to tick. The density of matter at that instant is 10^{94} grams per cubic centimeter[184] and the temperature is 10^{32} degrees Kelvin.

5.1.5. The four fundamental forces, quantum particle structure and grand unified theories

To understand how it is possible to compress all of the matter in the universe to a point of such density, we need to review briefly the four fundamental forces and the structure of quantum particles.

5.1.5.1. Four fundamental forces. The four fundamental forces of the universe are: gravity, the strong nuclear force, the weak nuclear force and electromagnetism. All four forces are the result of distortions in spacetime. The electromagnetic force governs the laws of chemistry and binds the atom together with the negatively charged electrons moving around the positively charged nucleus. This force is 10^{38} times stronger than the gravitational force but is effective over a very limited range. The electromagnetic force is about 10^{-2} of the strong force, but acts with a cumulative effect in the nucleus so that it competes with the strong force in determining the structure of the nucleus. The most common form of this force is light.

The force of gravity holds the solar system together and prevents the explosion of stars. Gravity keeps planets in their orbits and controls the movement of the galaxies. Although gravity is the weakest of the four forces and is of no significance in the subatomic world, it has infinite range and, unlike the nuclear or electromagnetic forces, cannot be negated or cancelled by another force. The gravitational force between two protons in a nucleus is only 10^{-38} of the strong force between them. The strength of gravity is related directly to the mass of the object causing the distortion in spacetime. Thus, the planet Jupiter causes more gravitational force than the planet Mercury because spacetime is less curved around Mercury than around Jupiter. (The force of gravity is more thoroughly explained in the discussion below on Einstein's theory of general relativity.)

The weak and strong nuclear forces overpower the electromagnetic force within the nucleus of the atom. The strong force operates in a short subatomic range and binds protons and neutrons together in the nucleus. Some particles, including the electron, do not feel the strong force; protons and electrons do not interact through the strong force. All protons in a nucleus carry positive charges. The strong force prevents the repulsive force between these charges from tearing the nucleus apart. When a nucleus has more than one hundred protons, the repulsive force between these protons is difficult for the strong force to overcome. The strong force makes a star bright, and as the star burns nuclear fuel, the energy of the strong force is released in the form of light. Obviously, with respect to our sun, this energy is necessary for the maintenance of life on earth.

The weak nuclear force is stronger than the gravitational force (about 10^{-7} of the strong force), but operates in a range much smaller than an atom's nucleus (smaller than 0.0001fm). This force controls the decay of a neutron into an electron, a proton and an antineutrino; and the activity of neutrons with other particles. The weak force is not a significant factor in the binding of nuclei.

5.1.5.2. Theory of quantum particle structure and quantum chromodynamics.

5.1.5.2.1. Overview. All matter is made up of atoms which consist of a nucleus surrounded by an electron cloud. At the atomic level the nucleus of the simple hydrogen atom consists of a single proton, a positively charged elementary particle. The nucleus of the deuterium form of hydrogen consists of a proton and a neutron, an elementary particle without an electrical charge. An electron carries a negative charge and has a mass of 9.11×10^{-31}kg. A proton has a mass 1836 times the mass of an electron. Protons and electrons interact with the electromagnetic force and are particles of matter known as fermions. The two categories of subatomic particles which make up matter and upon which the four fundamental forces act are quarks and leptons. Quarks are simple particles of matter which make up protons and neutrons. Leptons are simple particles of matter such as the electron or the neutrino. Quarks and leptons have mass but no dimensions. Because they have infinite density, they are called point masses.

The particles which transmit or carry the four forces are called bosons which have integer spins. Massless bosons convey the gravitational force (gravitons) and the electromagnetic force (photons), and massive bosons transmit the strong nuclear force (mesons and gluons) and the weak nuclear force (W and Z bosons). When electrons and protons exchange photons (the quanta of the electromagnetic force), the force is carried between the interacting particles. The force is not an invisible electron charge moving through space, but rather the absorption, exchange or emitting of photons. These photons, of course, travel at the speed of light.

Hadrons (including protons, neutrons, and pions) are a class of subatomic particles which interact through the strong force and are divided into baryons which decay into protons and mesons which decay into photons and leptons. Particles may be classified by their spin characteristics (see discussion below on the properties of spin). The proton, electron, and neutron all have spins of ½. Mesons all have integral spins (0, 1, 2, . . .). Baryons and all fermions have half-integral spins (1/2, 3/2, 5/2, . . .). Before moving deeper into the world of quarks, leptons, and bosons, it may assist our understanding if we make a brief digression into the world of antiparticles and then compare quantum structure with analogies between the atomic and subatomic world.

5.1.5.2.2. Antiparticles. Each particle has its antiparticle which is identical with the particle but has an opposite charge (except for particles with no charge such as the photon).[185] When particles collide with their antiparticles, they annihilate each other and transform into photons or energy and lighter particle-antiparticle pairs. Quarks and leptons cannot be destroyed or created individually but only in particle-antiparticle pairs. Antiparticles are very ephemeral because the collision between particles and antiparticles occurs quickly and, as the particles annihilate each other, energy is produced. Thus, a positron colliding with an electron transforms mass into energy. Consider the energy produced by a particle-antiparticle collision in the lepton family: an electron collides with its antiparticle, the positron. They annihilate each other and produce gamma radiation. From $E=mc^2$ we know that the amount of energy produced depends upon the mass of the electron which is 9.11×10^{-31}kg (the positron has the same mass). Inserting this amount of mass into the equation, we can calculate the amount of energy in the new gamma radiation as 1.02 MeV.[186] Because Einstein's famous equation works both ways, energy can also transform into matter. Gamma radiation equal to at least 1.02 MeV, under certain conditions, can transform into an electron-positron pair. Similarly, antiprotons and protons can be created but with higher energy requirements, because the mass of the proton is 1836 times as large as the mass of the electron.[187]

5.1.5.2.3. Analogies between atomic and subatomic world. Neil Bohr's concept of an atom with electrons orbiting around a nucleus has been replaced by abstract mathematical descriptions of electrons in a probability cloud enveloping a nucleus. At the atomic level particles are governed by the electromagnetic force and consist of electrons and protons which interact with the force and photons which carry the force. Neutrons can be part of the nucleus but do not interact with the electromagnetic force. This force holds the atom together by the attraction of the opposite electrical charges carried by these atomic particles.

Inside the nucleons (protons and neutrons) at the subatomic level are quarks which are analogous to electrons and protons as they also have electric charges, but are bound together by the strong nuclear force. At the subatomic level gluons play the role of photons by carrying the strong nuclear force which is also known as the color force. The color force has a color charge which is analogous to the electric charge of the proton and electron on the atomic level. As described more thoroughly below, these color charges come in three varieties: red, blue and green. The antiparticle partners of quarks are antiquarks which carry opposite color charges which are known as anti-red, anti-blue and anti-green. There are eight varieties of gluons which carry the strong force. Unlike the photons which are electrically neutral at the atomic level, at the subatomic level these gluons are also color charged. The attraction at the subatomic level is between opposite colors, such an anti-blue and blue. The interaction among similarly colored quarks is repulsion while quarks with colors which are not exact opposites attract but at a weaker intensity than opposite colored quarks. As discussed below, quarks must always combine so that particles are colorless or "white."

Particles have a property known as spin (intrinsic angular momentum). In quantum mechanics the electron rotates on its own axis with an angular momentum which is always ½h (where h is Planck's constant) or half as large as the smallest nonzero angular momentum in the motion of the electron around the proton and the hydrogen atom. This angular momentum is a vector quantity with direction and magnitude so that the value of the momentum will be either +½h or -½h. Accordingly, any electron can have a spin of +½h or -½h so that the electron can spin in one of two opposing directions.[188] The spin of a particle is similar to the side spin rotation of a basketball rebounding off a backboard. As a side spinning basketball will bounce off a backboard at a wider angle than a non-spinning ball, a particle with spin rebounds differently from one without spin. The Pauli principle (discussed below) requires that no two quarks or electrons with identical spins can occupy the same space. As Timothy Paul Smith has noted, time may be described as what keeps everything from happening all at once, and spin may be described as what keeps everything from happening in the same place.[189]

5.1.5.2.4. Leptons. As indicated above, leptons are point-like particles with no apparent internal structure which participate in the weak nuclear force but not in the strong nuclear force. Leptons come in six varieties: the electron, the electron neutrino, the muon, the muon neutrino, the tau and the tau neutrino. Neutrinos ("little neutral one" in Italian) have no electrical charge, have a spin of 1/2, and only interact with the weak force. The six varieties or flavors of leptons are divided into three families: tau, muon and electron. These three families are distinguished only by their masses. In terms of a ratio to the mass of a proton, their masses are as follows: 1/1836 for an electron, 1/9 for a muon, and 1/9 for a tau. The tau, muon, and electron all carry the same electrical charge.[190] Within the lepton category certain transitions among particles are possible. The following transitions can occur: a muon into a muon neutrino, a muon neutrino into a muon, a tau into a tau neutrino, a tau neutrino into a tau, an electron into an electron neutrino, and an electron neutrino into an electron. Transitions, however, cannot cross over the family lines. For example, a muon neutrino cannot turn into an electron or an electron turn into a tau neutrino.

5.1.5.2.5. Quarks.   Quarks are particles with fractional charges ($+\frac{2}{3}$ or $-\frac{1}{3}$) that form neutrons, protons, mesons, and other particles with mass. Mesons consist of a single quark and a single antiquark. Baryons consist of three quarks which are held together by gluons (as described below). Similar to leptons, quarks come in six flavors: up ($+\frac{2}{3}$ charge), down ($-\frac{1}{3}$), charmed ($+\frac{2}{3}$), strong ($-\frac{1}{3}$), top ($+\frac{2}{3}$) and bottom ($-\frac{1}{3}$). The proton is a baryon with three quarks ($+\frac{2}{3}, +\frac{2}{3}, -\frac{1}{3}$ = a positive 1 charge). The neutron also consists of three quarks ($+\frac{2}{3}, -\frac{1}{3}, -\frac{1}{3}$ = 0 charge). Each quark flavor has an equivalent antiquark with an opposite electric charge of the same value. Like leptons, quarks can make transitions within families but, unlike leptons, quarks can also make much weaker transitions to other quarks with different electrical charges.[191]

Austrian physicist Wolfgang Pauli established the principle that with respect to fermions, two identical particles cannot occupy the same quantum state (possess the same set of quantum numbers). Thus, two electrons cannot be in the exact same energy state, but can only be in the same energy level if they have opposite spins. The analogy between the subatomic model and the larger world appears in the statement that two identical things cannot be in the same place at the same time. To avoid this Pauli exclusion principle, a hypothesis concerning quarks proposes that quarks have properties of color charges. Each quark flavor occurs in three colors (red, green, and blue) and each antiquark occurs in three anticolors (yellow, magenta, and cyan). Thus, there are 18 different types of quarks and 18 different types of antiquarks. The gauge theory of color dynamics (quantum chromodynamics) requires that the color combinations must produce white in forming hadrons by combining red, green and blue to form baryons or combining one of these primary colors with its complementary anticolor to form mesons. The strong nuclear force binds the quarks and increases in strength as the distance between the quarks increases. This strong force which attracts and binds quarks occurs by the exchange of bosons known as gluons which are chargeless particles with zero rest mass. Gluons, in essence, "glue" quarks together as they convey the strong nuclear force. Each gluon carries a color charge and an anticolor charge. There are eight efficacious color-anticolor pairs because one of the pairs is equivalent to white and not involved in an exchange. A quark can alter its color charge in an interaction. Color charge alterations are attended by the emanation of a gluon. Another quark absorbs the gluon and changes its color charge to offset the original quark's color alteration. Thus, if a blue quark changes to red, the emitted gluon will have blue and anti-red color charges. When this gluon is absorbed by a red quark, the red color charge of the quark and the anti-red color charge of the gluon annihilate each other which leaves the second quark with a blue color charge. Notice that the net effect of these changes are offsetting in that the original blue quark and red quark have been transformed into a red quark and a blue quark. Thus hadrons are always white even though the quark colors change as the gluons convey the strong nuclear force among quarks.[192]

5.1.5.2.6. Dimensionless features of quarks and leptons, the singularity, and the excess of particles over antiparticles. In the context of this background on subatomic particles, the possibility of the compression of all of the universe's matter into one point remains because quarks and leptons have no dimensions. Consequently, the universe could be compressed into a single point of infinite density. This compression would produce such intense heat that all quantum particles would cease to exist. As stated, space and time would no longer exist, because they would be infinitely distorted or curved around this mass. Under the Big Bang theory time and space would not exist outside this singular point. At the infinite or near infinite intense heat at this singular point, Einstein's famous equation $E = mc^2$ means that matter and energy are interchangeable. "Prior" to Planck time only energy existed.[193] The net effect of a transformation from energy to mass created no electric charge changes. The number of photons, protons, antiprotons, electrons, and positrons remained equal. According to the hypothesis, the Big Bang explosion ultimately resulted in a

slight excess of protons and electrons over positrons and antiprotons. This slight excess made up the matter in the universe.[194]

5.1.5.3. Grand unified extra dimensional theories.

5.1.5.3.1. Guts and strings.   Physicists are still searching for a
grand unified theory or GUT, which would explain a unity
among the four fundamental forces. One of the most promising
theories is string theory which proposes that the components of
subatomic particles are not pointlike masses but tiny vibrating
strings. The difference in vibration (like the difference in the vibration of a stringed musical instrument) determines the properties of a subatomic particle. The superstring theory requires ten
dimensions (one of time and nine of space). The theory proposes
that ten dimensions existed in the first instant of the Big Bang.
Four flattened out and six compacted and became invisible existing only at the very subatomic level. To appreciate the characteristics of strings and the promise the theory holds in explaining a
vast variety of physical phenomena, a brief digression into the
history of the concepts of symmetry, supersymmetry, supergravity, and superstrings may be informative.

*5.1.5.3.2. Symmetry and Kaluza-Klein extra dimensional
theories.*   The idea of symmetry is simply that different physical phenomena function pursuant to an underlying basis. In the
nineteenth century, Michael Faraday discovered the relation between electricity and magnetism which Scottish physicist James
Clerk Maxwell later quantified in his equations as a united electromagnetic force. In 1919, Theodore Franz Éduard Kaluza, a
German mathematician, wrote a letter to Albert Einstein in
which Kaluza proposed a fifth dimension in order to establish a
unified basis for gravity and electromagnetism, the two forces of
nature known at that time. Kaluza postulated this extra dimension as a very small, unobservable, curled up loop existing in every location of ordinary space. When working in an extra
dimension a mathematician merely adds on extra terms to an
equation so that mathematical thought processes can take into
account five or fifty or even one hundred dimensions. For example, Einstein wrote his equations for the theory of general

relativity in four dimensions. When Kaluza added extra terms to the equation for the fifth dimension, his equations produced the theory of general relativity and the laws of electromagnetic interactions.

In 1926 Swedish physicist Oskar Klein reconciled Kaluza's theory with quantum mechanics and realized that subatomic particles related to the peculiar vibrations of the compact looped strings. In Kaluza's equations of five dimensional gravity the diameter of the compacted fifth dimension is approximately 10^{-32} meters, an incredibly small distance. As indicated above, at the time of Kaluza and Klein's writings, only two fundamental forces were known. Today we know of the strong and weak nuclear forces, so modern Kaluza-Klein theories attempting to produce a unified basis for all four fundamental forces must have more than five dimensions. Initially it appeared that the optimum number of dimensions was eleven so that at every point in ordinary space and at every moment of time, an extremely compact seven dimensional structure exists which is too small to be observed with even the most powerful microscope.[195]

5.1.5.3.3. Supersymmetry and supergravity theories. This conclusion is supported by the theory of supersymmetry combined with general relativity. Supersymmetry is a symmetry that can be applied to bridge the divide between bosons with whole number amounts of spins and fermions. According to supersymmetry theories, every boson has a corresponding fermion partner and vice versa. The boson partners of fermions have names derived by adding "s" to the beginning of the name of the fermion. Thus, the electron has a supersymmetric partner known as the selectron; the quark has a supersymmetric partner known as the squark; the neutrino has a supersymmetric partner known as the sneutrino; and the lepton has a supersymmetric partner known as the slepton. Similarly, the fermion partners have names derived by replacing the "on" ending of the boson's name with the letters "ino." Thus, the graviton has a supersymmetric partner known as the gravitino; the photon has a supersymmetric partner known as the photino; and the gluon has a supersymmetric partner known as the gluino.

The combination of supersymmetry theories with general relativity produced the supergrand unified theory known as supergravity. For some time, this became the favorite grand unified theory of Murray Gell-Mann and Stephen Hawking. These physicists were so enthusiastic about supergravity that Hawking pronounced that the theory could mean an end to theoretical physics. He was wrong. The supergravity theories ran into several intractable mathematical problems. They were not consistent with our knowledge of leptons, quarks, and W particles. The theories could only solve a technical property of fermions when an even number of dimensions were used.[196] The equations of supergravity theories also produced infinities when particles were compacted together as they were during Planck time.

5.1.5.3.4. String theories. Physicists turned their attention to a more promising theory resulting from the work of John Schwarz of the California Institute of Technology and Michael Green, who expanded on a 1960's paper by Gabriele Veneziano. In that paper, Veneziano developed a formula which proposed that hadrons are comprised of tiny strings which produce different properties of matter by vibrating in different modes.[197] Schwarz and Green combined Veneziano's string theory with supersymmetry to produce the concept of superstrings which function as tiny loops of energy. ("Super" is added as a prefix to strings as a sign that the theory has supersymmetry). This concept removed the problem of infinities in the mathematical equations and calculated the length of superstrings as equal to the Planck length or 10^{-35} meters.

Schwarz and Green's work became the basis for the work of Princeton physicist Edward Witten, currently at the Institute of Advanced Study at Princeton, and the String Quartet, a group of physicists which produced a mathematical formula for the heterotic string which in its loop form combined a ten dimensional string theory associated with fermions with the older twenty-six dimensional string theory associated with bosons. Under the heterotic superstring theory six of the twenty-six dimensions are compacted and so small that they are unobservable. Four dimensions (length, width, height and time) are the flattened out dimensions of the universe, and sixteen dimensions are "internal dimensions" which work in the heterotic strings to account for the four fundamental forces.[198] These forces can be unified by abstract mathematics in a ten dimensional universe which is required by this theory. Under the theory, quantum mechanics and special and general relativity are consistent; the theory results in spin 2 particles which are gravitons and consequently contains a quantum theory of gravity. In other words, superstring theory yields the theories of special and general relativity and is consistent with the principles of quantum physics. The theory is actually a quantum theory which requires gravity. Although still speculative, superstring theory has many impressive consistencies. As Hugh Ross notes:

> It is the only theory that self-consistently explains all the known properties of the known fundamental particles (now numbering fifty-eight), all the properties and principles of quantum mechanics, all the properties and principles of both special and general relativity, the operation of all four forces of physics, and all the known details of the creation event.[199]

Theorists believe that vibrating superstrings were greatly stretched at the first instant of the existence of the universe when the universe had ten dimensions and contracted as temperatures lowered to such a degree that the strings now act like points. At such a short length of 10^{-35} meters, the related energy level would be about 10^{19} GeV, a level outside our ability to ever duplicate with an accelerator. As indicated above, the superstring theory postulates that the universe split into a four dimensional sector which expanded and produced the universe we observe and a six dimensional section which curled up or compacted in an invisible structure that exists everywhere at every location, *hic et ubique*,[200] within the four dimensions. For the purpose of the formation of life, this split was fortunate, because carbon-based life could not exist in any other than three spatial dimensions. Gravity would not allow for stable planetary systems unless it functioned in three spatial dimensions, because it follows an inverse square law which requires the force of gravity to decrease as distance increases. In four spatial dimensions, the force of gravity would fall to a fraction of one-eighth its power (rather than one-quarter) for every doubling of distance, and in five spatial dimensions, the force would fall to one-sixteenth for every doubling of distance. Moreover, in more than three spatial dimensions, the force of electromagnetism would not function in a manner which would allow for life, because electrons would either spiral away from or into the nuclei. Neutral atoms and molecules could not exist, nor could stable stars.[201]

Until recently, string theory was plagued by its ten dimensional equations having tens of thousands of six dimensional solutions with at least that many four dimensional models. In 1995, Andrew Strominger of the University of California, Brian Greene of Cornell University and David Morrison of Duke University, reported that when string theory takes into account the quantum effects of time with charged extremal black holes having a mass equal to an elementary particle, the myriad of solutions become only one. Strominger has said that the equations indicate that strings and these simplest, smallest black holes (which are the proposed destiny of large decaying black holes) are actually two descriptions of the same thing. Black holes could turn into strings and vice versa. As Gary Taube writes:

> In string theory . . . the six extra dimensions of the 10-dimensional theory curl up into structures known as Calabi-Yau spaces, which are the ones that regrettably seem to come in tens of thousands of possible configurations. Now Greene, Morrison, and Strominger have shown that during the phase transition, the Calabi-Yau spaces would evolve into one another, and at the same time black holes would become strings and vice versa. These transformations not only imply that black holes and strings are two different descriptions of the same fundamental object, but also that there may not be tens of thousands of four-dimensional solutions to the equations after all.[202]

The direction of string theory appears to be heading towards one solution. The five different kinds of strings (heterotic and type II strings) envisioned by theorists are being integrated into a single theory. Two of the leading authorities on string theory, Edward Witten and John Schwarz voice their optimism:

> For starters, says Witten, what once appeared to be five distinct string theories, "we now know are all equivalent to each other." Adds Caltech theorist, John Schwarz, one of the originators of modern string theory: "It's now clear there is just one string theory, which is nice, because it's the only known theory consistent with gravity and quantum mechanics. We only need one. Now the claim is we only really have one." (Theorists) now have the tools, they said, to explore versions of their theories with less supersymmetry and greater realism. "We're marching away from the highly symmetric configurations," says Strominger, "and marching toward situations in which there's less and less symmetry, and we're marching full steam, and we haven't run into a road block yet."[203]

Although experimental verification of this new theory may be many years away, it does not suffer from some of the flaws of its failed predecessors.[204] A number of physicists, including Hugh Ross, consider string theory to be useful in resolving many metaphysical and theological paradoxes, including determinism and free will and the transformation of a rectangle into a point. For the purposes of the first question presented in this book, we need merely note that string theory appears to open an elegant and aesthetic window to a mosaic of unusual precision and intricacy.

5.1.6. Particle and fundamental force activity in the early universe

Ad initium, the universe began with the formation of space, time, energy, and matter. All four fundamental forces were unified in a single force. The event horizon for the universe was Planck time (10^{-43}s) with the temperature of the universe at 10^{32} degrees Kelvin. At Planck time gravity separated from the unified force. According to one model, the universe's temperature was 10^{27} degrees K at 10^{-35}s when the universe inflated at a speed exceeding the speed of light. During this inflationary epoch the universe expanded by a factor of 10^{50} until the epoch ended at 10^{-33}s. Although this expansion exceeded the speed of light, it did not violate Einstein's principle that nothing can travel in space faster than the speed of light. This expansion was the expansion of space itself and not the motion of particles through space. Space expanded at a speed exceeding 2.998×10^8 meters per second; the objects did not move through space at a speed greater than the speed of light.

After the inflationary epoch, particles and antiparticles destroyed each other with a small amount of matter remaining. At 10^{-12}s the temperature of the universe fell to 10^{15} degrees K, and the weak and electromagnetic forces separated. At 10^{-6}s, antiquarks and quarks stopped destroying each other and protons and neutrons were formed. At 10^{-4}s, the capture of electrons and positrons caused protons and neutrons to exchange identities so that neutrons became protons and protons became neutrons. Because more energy is required for the formation of neutrons, protons outnumbered neutrons by a factor of five. At 1 second neutrinos no longer interacted with the primordial fireball, neutrinos decoupled, and the universe became transparent to neutrinos.

During the first seconds of the universe, the temperature was several billions of degrees Kelvin and too high for nuclei to stick together. Photons (radiation) outnumbered neutrinos, protons, neutrons and electrons (particles) so that the interaction of photons and particles was controlled by radiation and the universe was radiation dominated. Protons, positrons, neutrons, neutrinos, electrons, antineutrinos, and photons all were in a state of thermal equilibrium. After the first minute nuclear reactions began to occur, and protons and neutrons interacted to make heavy hydrogen or deuterium which could capture an additional proton to form helium-3 or capture another neutron to form tritium. Helium-4 could then be formed, removing most available neutrons so that 75% of the universe consisted of hydrogen and approximately 25% consisted of helium. At this time primordial cosmic nucleosynthesis formed the light elements. After four minutes the feverish nuclear reactions of nucleosynthesis stopped and the opaque radiation dominated universe continued to expand and cool with electrons at an energy level too high to interact with the nuclei of atoms.

After about 10,000 years, the temperature of the expanding universe had fallen so that the radiation wavelengths had lengthened to emit ultraviolet waves. As the wavelengths were stretched, the photons lost energy.[205] Pursuant to the equation $E = mc^2$, the energy density of matter is inversely proportional to the mass or number of particles in the universe. As the universe continued to expand and photons continued to lose energy, the mass density of particles exceeded the mass density of photons and the universe became matter dominated. The temperature of the universe continued to fall as the universe continued to expand and radiation wavelengths continued to increase with a loss of radiation energy until the energy from radiation was no longer strong enough to prevent electrons from interacting with protons and forming stable hydrogen atoms. This is the time of recombination and decoupling when the temperature of the universe had fallen to 3,000°K, approximately 300,000 years after the Big Bang. The disappearance of free electrons and the lack of the availability of these electrons to interact with protons caused the end of thermal equilibrium and the separation of matter and radiation. The cosmic background radiation measured precisely by the COBE satellite at 2.726°K is the red-shifted remnant from the time when radiation and matter were last in thermal equilibrium. After about 1 million years, all of the protons and electrons had combined into hydrogen atoms, and photons were able to travel freely. At this time the universe emerged from its dense opaque fog and became transparent. After the universe became transparent, the temperature dropped quickly, and radiation cooled proportionately with the universe's expansion rate to the present 2.726°K.

In the early universe, atoms formed a uniformly dense gas. As the universe expanded, the gravitational force caused local condensations in this gas. This condensation later combined masses of matter into large bodies which experienced internal condensation which formed galaxies. From these galaxies clusters appeared and finally stars appeared. After approximately 1 billion years, quasars formed. After about 7.5 billion years our sun, earth, and solar system emerged out of the Milky Way galaxy.

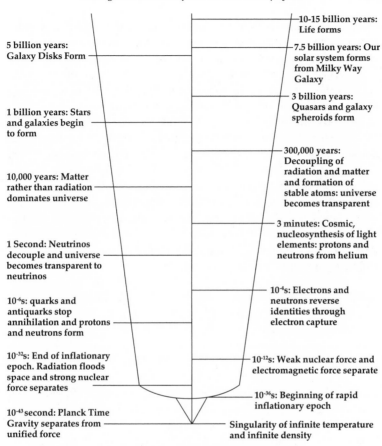

5 billion years:
Galaxy Disks Form

1 billion years: Stars
and galaxies begin
to form

10,000 years: Matter
rather than radiation
dominates universe

1 Second: Neutrinos
decouple and universe
becomes transparent to
neutrinos

10^{-6}s: quarks and
antiquarks stop
annihilation and protons
and neutrons form

10^{-32}s: End of inflationary
epoch. Radiation floods
space and strong nuclear
force separates

10^{-43} second: Planck Time
Gravity separates from
unified force

10-15 billion years:
Life forms

7.5 billion years: Our
solar system forms
from Milky Way
Galaxy

3 billion years:
Quasars and galaxy
spheroids form

300,000 years:
Decoupling of
radiation and matter
and formation of
stable atoms: universe
becomes transparent

3 minutes: Cosmic,
nucleosynthesis of light
elements: protons and
neutrons from helium

10^{-4}s: Electrons and
neutrons reverse
identities through
electron capture

10^{-12}s: Weak nuclear force and
electromagnetic force separate

10^{-36}s: Beginning of rapid
inflationary epoch

Singularity of infinite temperature
and infinite density

BIG BANG (INFLATIONARY MODEL)
TIME TABLE

[Figure 4]

5.2 Examples of precision of values in particle astrophysics necessary for life

With this background in particle astrophysics and the activity of the forces and particles in the early universe, we are now in a position to turn to examples of the fine tuning of the values in particle astrophysics which are a prerequisite for the formation of life. In reviewing the physical laws and the numerical values of fundamental constants, one encounters a remarkable precision in these values such that only small changes in the fundamental constants, such as the strength of the four forces, Planck's constant, the mass of elementary particles, etc., would yield a universe without galaxies, stars, atoms or even nuclei, and consequently, without the capacity for life.[206] Many physicists have compiled lists of "cosmic coincidences,"[207] "just right" characteristics of the universe,[208] and "unnatural selections."[209] The constants of nature, such as the strength of the gravitational force, have exactly the values necessary for the existence of stars and planets. Although one could make an extremely long compilation of finely tuned values, the following examples will provide a glimpse of the mathematical probabilities which argue against accident as an explanation for the formation of a universe compossible with life.

5.2.1. *Resonance precision required for existence of carbon, a necessary element for life*

Every living cell contains proteins which function as enzymes that act as catalysts in cell reactions. Proteins are made from the linkage of amino acids which all have a central carbon atom. Thus carbon is essential for life. The carbon atom is the fourth most common element in our galaxy. Life would be impossible without carbon and yet because of the precise requirements for its existence, the carbon atom should be very rare. The formation of a carbon atom requires a rare triple collision known as the triple alpha process. The first colliding step in this process occurs when a helium nucleus collides with another helium nucleus within a star. This collision produces an unstable, very ephemeral isotope of beryllium known as Be^8 (Be^9 is beryllium's stable form). When the unstable, short lived beryllium collides with a third helium nucleus, a carbon nucleus is formed.[210]

Astrophysicist Sir Fred Hoyle predicted the resonances (energy levels) of the carbon and oxygen atoms. The resonance of the carbon nucleus is precisely the right resonance to enable the components to hold together rather than disperse. This resonance perfectly matches the combined resonance of the third helium nucleus and the beryllium atom. Owen Gingerich, Professor of Astronomy and of the History of Science at the Harvard-Smithsonian Center for Astrophysics, discusses resonance generally and the precise resonance within carbon that assists the triple alpha process:

> As you tune your radio, there are certain frequencies where the circuit has just the right resonance and you lock onto a station. The internal structure of an atomic nucleus is something like that, with specific energy or resonance levels. If two nuclear fragments collide with a resulting energy that just matches a resonance level, they will tend to stick and form a stable nucleus. Behold! Cosmic alchemy will occur! In the carbon atom, the resonance just happens to match the combined energy of the beryllium atom and a colliding helium nucleus. Without it, there would be relatively few carbon atoms. Similarly, the internal details of the oxygen nucleus play a critical role. Oxygen can be formed by combining helium and carbon nuclei, but the corresponding resonance level in the oxygen nucleus is half a percent too low for the combination to stay together easily. Had the resonance level in the carbon been 4 percent lower, there would be essentially no carbon. Had that level in the oxygen been only half a percent higher, virtually all of the carbon would have been converted to oxygen. Without that carbon abundance, neither you nor I would be here.[211]

By his own admission, Hoyle's atheism was dramatically disturbed when he calculated the odds against these precisely matched resonances existing by chance. Hoyle wrote:

> A common sense interpretation of the facts suggests that a superintellect has monkeyed with physics, as well as with chemistry and biology, and that there are no blind forces worth speaking about in nature. The

numbers one calculates from the facts seem to me so overwhelming as to put this conclusion almost beyond question.[212]

5.2.2. *Explosive power of Big Bang precisely matched to power of gravity; density precisely matched with critical density*

For the universe to form, the force of gravity had to match precisely the explosive force of the Big Bang. If the force of explosion was only slightly higher, the universe would only consist of gas without stars, galaxies or planets. Without stars, galaxies and planets, life could not exist. The matching had to be to the remarkable precision of one part in 10^{55}.[213] If the rate of expansion was reduced by only one part in a thousand billion, the matter in the universe would have collapsed back to a singular point after a few million years. In his book, *In the Center of Immensities*, physicist Bernard Lovell wrote about the extraordinary precision in this essential condition for life and wondered about the significance of such precision:

> We have attempted to describe the early stages of the expansion of the universe but the *description* in terms of nuclear physics and relativity is not an *explanation* of those conditions. Formidable questions arise and it is not clear today where the answers should be sought: indeed, even the scientific description of these queries produces the remarkable idea that there may not be a solution in the language of science. Why is the universe expanding? Furthermore, why is it expanding at so near the critical rate to prevent its collapse? The query is most important because minor differences near time zero would have made human existence impossible. When the universe was one second from the beginning of the expansion we have stated that the temperature had fallen to 10^{10} deg K and the density to 1 gram per cubic centimeter. It is a phase when, it is postulated, the universe had already reached the possibility of description in terms of common physical concepts. If at that moment the rate of

expansion had been reduced by only one part in a thousand billion, then the universe would have collapsed after a few million years, near the end of the epoch we now recognize as the radiation era, or the primordial fireball, before matter and radiation had become decoupled. This remarkable fact was pointed out recently by one of the most distinguished contemporary cosmologists who referred to the suggestions that out of all the possible universes the only one which can exist, in the sense that it can be known, is simply the one which satisfies the narrow conditions necessary for the development of intelligent life.[214]

Paul Davies is also impressed with the degree of precision in matching the explosive force and the force of gravity. Davies wrote about the sensitivity of this rate in the very first seconds of the universe's existence. His calculations concluded that at Planck time the matching was precise to an astounding one part in 10^{60}. If the explosion had varied in strength by only one part in 10^{60}, a universe compossible with life would not exist. In his words: "To give some meaning to these numbers, suppose you wanted to fire a bullet at a one-inch target on the other side of the observable universe, twenty billion light years away. Your aim would have to be accurate to that same part in 10^{60}."[215]

Closely related to the fine tuning of the expansion rate is the density of the universe. The rate of expansion in the universe decelerates over time because of the gravitational forces of the galaxies. The rate of deceleration depends upon the density of the matter in the universe. At the critical density, the universe has almost no curvature and is flat. The critical density of matter in the universe is the average density which provides for flat space.[216] If the density of matter (P) after Planck time were slightly greater than the critical density (P_c), the universe would have rapidly collapsed back into an infinitely dense singularity. If P had been only slightly less than P_c the universe would have expanded so that no stars and consequently no life could have formed. With only the very slightest deviation from $P = P_c$, life would have been impossible. William J. Kaufmann III has commented on the extraordinary precision in this value:

Consequently, the density of the universe immediately after the Big Bang must have been equal to the critical density to an incredibly high order of precision. Calculations demonstrate that, in order for the rate of deceleration to be roughly 1 today, the value of P right after the Big Bang must have been equal to P_c to more than 50 decimal places! What could have happened immediately after the Planck time to ensure that $P=P_c$ to such an astounding degree of accuracy?[217]

Some physicists believe that one explanation can be found in a model of an inflationary epoch at about 10^{-35} of the first second where a short period of accelerated expansion caused the perfect balance between gravity and the rate of expansion and density and critical density. This could explain the very flat characteristics of the universe given by these precise matchings, but the inflation required in this model would itself require an extraordinary fine tuning to yield the precisely balanced result. If the inflationary model is true, the inflationary epoch would contain enormous fine tuning and the precision of values issue is only removed one step by the inflationary model. This model does not explain why the inflationary epoch (if it occurred) was so finely tuned to produce such a staggering degree of balance.

5.2.3. Delicate balance in strong nuclear force

The strong nuclear force binds the particles in an atom's nucleus and is the strongest of the forces, approximately one hundred times as strong as the electromagnetic force which in turn is ten thousand times stronger than the weak nuclear force. The weak nuclear force is approximately ten thousand billion, billion, billion times stronger than the force of gravity. Considering these enormous differences in strength, one can appreciate the remarkable precision required to balance these forces to a degree of one part in 10^{60}.[218]

If the strong nuclear force were any weaker, nuclei of atoms would not hold together. The universe would only have one element: hydrogen; deuterium (hydrogen atoms with a nuclei consisting of a proton and a neutron) would not exist. Any reduction in the strong nuclear force would dissolve the bond between the proton and neutron in deuterium. Deuterium is a crucial ingredient in the nuclear reaction which keeps stars like the sun burning and thus is an element required for the existence of life.

If the strong nuclear force were only 2 percent stronger, two protons could bind despite their electrically charged repulsions, and hydrogen would be an unusual element in the universe. The universe would consist mainly of helium and very little hydrogen. Hydrogen, of course, was necessary for the formation of the sun and the formation of liquified water, both necessary for the formation of life. The existence of combined protons would also make hydrogen explosive to a catastrophic extent.

A 2 percent increase in the strength of this force would not only cause a lack of hydrogen, but a lack of the heavier elements necessary for life. Such an increase would prevent quarks from forming protons. A 2 percent decrease in the strength of this force, on the other hand, would render unstable certain of the heavy elements which are prerequisites for life. The weak force controls the sun's burning of hydrogen in a slow and steady manner. Diprotons would produce deuterium which would place this burning under the control of the strong nuclear force which would burn hydrogen at a rate 10^{18} times faster than the burning under the control of the weak nuclear force. This would cause a debacle of hydrogen consumption and would consume most if not all hydrogen in the hot primeval phase, leaving helium as the only element in the universe.[219] In commenting on the precise balance between the strong and weak nuclear forces, John Polkinghorne writes:

> In the first three minutes of cosmic history, the whole universe was the arena of nuclear reactions. When that era came to an end, through the cooling produced by expansion, the world was left, as it is today on the large scale, a mixture of three-quarters hydrogen and one-quarter helium. A little change in the balance between the strong and weak nuclear forces could have resulted in there being no hydrogen—and so ultimately no water, that fluid that seems so essen-

tial to life. A small increase (about 2 percent) in the
strong nuclear force would bind two protons to form
diprotons. There would then be no hydrogen - burning main - sequence stars, but only helium burners,
which are far too fierce and rapid to be energy sources capable of sustaining the coming to be of planetary
life. A decrease in the strong nuclear force by a similar
amount would have unbound the deuteron and
played havoc with fruitful nuclear physics.[220]

5.2.4. *Balancing of gravitational force and electromagnetic force*

The strength of the force of gravity precisely matches the
strength of the electromagnetic force to allow for the formation of
a star such as the sun. The gravitational force holds the star together and contributes to the amount of pressure inside the star.
Electromagnetism produces the radiation which is the flow of energy out of the star. The surface temperature of a star is related to
the central temperature of its core through its luminosity which is
defined as the flow of energy (radiation) per unit of time into
space. If a star is static and in a steady state, then the rate of loss
of energy from the surface is precisely balanced by the rate of energy generation in the core. The rate of energy generation in the
core is controlled by the central temperature. The central temperature will adjust automatically so that the rate of production of
energy from the nuclear interior burning is equal to the flow of
energy from the surface.

For a star to avoid convective instability its interior must be
able to radiate energy fast enough to avoid convection which
causes the surface to boil. This convection is seen in lighter stars
which are cool and convective and known as red dwarfs. Heavier
stars divest themselves of energy generated in their core in the
form of radiation and light. These hotter stars, known as blue giants, do not need convection to assist in the transportation of energy from the core. Most stars fall in the narrow range between
blue giants and red dwarfs. If the force of gravity, however, was
changed by only one part in 10^{40}, all stars would be either blue giants or red dwarfs. Stars, like the sun, would not exist nor would

any life dependent on such a star. If the electromagnetic force were only slightly stronger, all stars would be red and too cold for the emergence of life. If the force of electromagnetism were slightly weaker, all stars would be extremely hot blue giants which would burn out before life could emerge from any planet orbiting a star. As Paul Davies writes in *The Accidental Universe*:

> Nature has evidently picked the values of the fundamental constants in such a way that typical stars lie very close indeed to the boundary of convective instability. . . . If gravity were *very* slightly weaker, or electromagnetism *very* slightly stronger (or the electron slightly less massive relative to the proton), all stars would be red dwarfs. A correspondingly tiny change the other way, and they would all be blue giants. . . . a star's convection plays an important role in planetary formation, so that a world where gravity was very slightly less weak might have no planets. In either case, weaker or stronger, the nature of the universe would be radically different.[221]

John Leslie gives the following description concerning the balance between gravity and the electromagnetic force:

> While the figure varies with whether we consider electron-electron or proton-proton interactions, we can say roughly that gravity is an astonishing 10^{39} times weaker than electromagnetism. Were it appreciably *stronger* than it is, stars would form from small amounts of gas; and/or they would blaze more fiercely (E. Teller calculated in 1948 that stellar radiation would increase as the seventh power of the gravitational constant, and in 1957 Dicke linked this to how a change making gravity slightly nearer in strength to electromagnetism would mean that long ago all stars would be cold. This would preclude the existence of man); and/or they would collapse more easily to form white dwarfs, neutron stars, or black holes."[222]

5.2.5. Meticulous balance between number of electrons and protons

For the existence of life, the number of electrons must be meticulously balanced to an accuracy of one part in 10^{37} with the number of protons. Without this balance the force of gravity which was essential to the formation of stars and planets would have been overwhelmed by the electromagnetic force. Hugh Ross notes that 10^{37} is difficult to comprehend and gives the following visual analogy to demonstrate the precision of the value:

> Cover the entire North American continent in dimes all the way up to the moon, a height of about 239,000 miles. (In comparison, the money to pay for the U.S. federal government debt would cover one square mile less than two feet deep with dimes.) Next, pile dimes from here to the moon on a billion other continents the same size as North America. Paint one dime red and mix it into the billion piles of dimes. Blindfold a friend and ask him to pick out one dime. The odds that he will pick the red dime are one in 10^{37}. And this is only one of the parameters that is so delicately balanced to allow life to form.[223]

5.2.6. Precision in electromagnetic force and ratio of proton mass to electron mass and neutron mass to proton mass

The electromagnetic force binds protons and electrons in atoms. An electron's path around an atom's nucleus governs the ability of the atom to bond with another atom in the formation of molecules. If the electromagnetic force were slightly increased in strength, an atom would not share an electron with other atoms and molecules would not form. If the force were slightly weaker, the electrons would not remain in their paths around an atom's nuclei. Accordingly, any change in the strength of the electromagnetic force would preclude the formation of life.[224] The electromagnetic force must be precisely balanced with the ratio of electron mass to proton mass. If the ratio of the electron mass to proton mass were not precisely balanced, the chemical bonding required for life would not occur.

The proton is 1,836 times heavier than the electron. This is a fundamental number in nature, finely tuned to allow for the formation of molecules. Stephen Hawking comments on the precision of the ratio:

> The laws of science, as we know them at present, contain many fundamental numbers, like the size of the electric charge of the electron and the ratio of the masses of the proton and the electron. . . . The remarkable fact is that the values of these numbers seem to have been very finely adjusted to make possible the development of life. For example, if the electric charge of the electron had been only slightly different, stars either would have been unable to burn hydrogen and helium, or else they would not have exploded.[225]

Similarly, another crucial ratio for the emergence of life is the precise difference between the mass of the neutron and the mass of the proton. If these masses were not about double the electron's mass, stable nucleides would not exist. Nucleides make up the elements which are required for the reactions of chemistry which are a prerequisite for life. The precision in the differential between the mass of the proton and the mass of the neutron, however, allowed for a universe meticulously balanced with one neutron for every seven protons. The neutron is more massive than the proton by about one part in a thousand. The proton has less energy so if the difference were greater, neutrons would decay into protons and the force of electromagnetism would then blow apart the nuclei. This would yield a universe of only protons with hydrogen the only possible element in the universe. Neutrons are required to form all the other elements because they bring the strong nuclear force to hold nuclei together without bringing sufficient electromagnetic repulsion to cause their disintegration. If there were a slightly smaller difference among these masses, free neutrons would not decay into protons and hydrogen would not exist.[226] The emergence of life depended on the precision among these three masses with the proton mass at 938.28 MeV, the electron at .51 MeV (for a total of 938.79 MeV) and the neutron at 939.57 MeV. John Leslie comments on this delicate balance which allows for the emergence of life in the universe:

> . . . the Big Bang cooled just quickly enough to allow neutrons to become bound to protons inside atoms. Here the presence of electrons and the Pauli principle discouraged their decay, but even that would not prevent it were the mass difference slightly greater. And were it *smaller*—one third of what it is—then neutrons *outside* atoms would *not* decay. All protons would thus change irreversibly into neutrons during the Bang, whose violence produced frequent proton-to-neutron conversions. There could be no atoms: the universe would be neutron stars and black holes . . . *The mass of the electron* enters the picture like this. If the neutron mass failed to exceed the proton mass by a little more than the electron mass then atoms would collapse, their electrons combining with their protons to yield neutrons. . . . As things are, the neutron is just enough heavier to ensure that the Bang yielded only about one neutron to every seven protons. The excess protons were available for making the hydrogen of long-lived stable stars, water, and carbohydrates.[227]

5.2.7. Big Bang's defiance of Second Law of Thermodynamics and gravity's cumulative effect

Thermodynamics is the study of the interrelation between heat and other forms of energy. The First Law of Thermodynamics states that energy and matter can neither be created nor destroyed. The Second Law of Thermodynamics requires that entropy or disorder in the universe tends toward a maximum. The contents of the universe are becoming less ordered, and as the universe becomes more disorganized, less of its energy is available to perform work. Because the universe is running down, it must have had a beginning. The universe could not be dissipating from infinity. Reversing the observed process of dissipation, the Second Law of Thermodynamics requires a beginning and a very highly ordered beginning (one with low entropy). If the Big

Bang is regarded as only a big, impressive accident, there is no explanation why the Big Bang produced a universe with such a high degree of order, contrary to the Second Law of Thermodynamics, especially considering the cumulative power of gravitating systems in the universe. When quantum theory is applied to black holes, the formula for the entropy of such a singularity is vastly greater than for the entropy of a star of the same mass. The black hole appears to represent the equilibrium end state of a gravitating system, the state at which the system reaches maximum entropy. Note in the following comment by Paul Davies that in a 1979 calculation *Roger Penrose computed that the probability of the observed universe occurring by chance is one in 10^{300}*:

> Given a random distribution of (gravitating) matter, it is overwhelmingly more probable that it will form a black hole rather than a star or a cloud of dispersed gas. These considerations give a new slant, therefore, to the question of whether the universe was created in an ordered or disordered state. If the initial state were chosen at random, it seems exceedingly probable that the big bang would have coughed out black holes rather than dispersed gases. The present arrangement of matter and energy, with matter spread thinly at relatively low density, in the form of stars and gas clouds would, apparently, only result from a very special choice of initial conditions. Roger Penrose has computed the odds against the observed universe appearing by an accident, given that a black hole cosmos is so much more likely on *a priori* grounds. He estimates a figure of 10^{300} to one.[228]

The figure for the total entropy of the Big Crunch (the reversal of the Big Bang into a singularity) calculated by Roger Penrose at a volume of 10^{123} allows one to estimate the total phase-space volume available for the precision required in the Big Bang to begin the universe in a highly ordered state consistent with the Second Law of Thermodynamics. (Penrose's calculations are also discussed in the section of this book concerned with the case against an oscillating universe.) According to a 1989 calculation of

Penrose, the entropy of 10^{123} should represent the largest phase-space volume available for the resulting precision. 10^{123} is the logarithm of this volume so that the volume would be the exponent 10^{123}. Penrose answers the question how precise the Creator's aim had to be in order to provide for a universe compatible with the Second Law of Thermodynamics:

> This now tells us how precise the Creator's aim must have been: namely to an accuracy of
>
> one part in $10^{10(123)}$.
>
> This is an extraordinary figure. One could not possibly even *write the number down* in full, in the ordinary denary notation: it would be "1" followed by 10^{123} successive "0"s! Even if we were to write a "0" on each separate proton and on each separate neutron in the entire universe—and we could throw in all the other particles as well for good measure—we should fall far short of writing down the figure needed.[229]

Focus on the number $10^{10(123)}$ for a moment. If one were to write a "1" and then a zero on every atomic particle (not just every atom, but every atomic particle within the atom) in this planet, one would not be able to write the number down. If a zero was written on every atomic particle in the solar system, one could not write the number down. If a zero was written on every atomic particle in the Milky Way galaxy, one could not write the number down. If a zero was written on every atomic particle in the observable universe, one would still fall far short of the matter necessary to even write the number down. The precision needed in order for our universe to exist is so extraordinary that there is not enough matter in the known universe for written numbers to describe it in the ordinary denary notation.

5.2.8. Delicate balance of values related to weak nuclear force

The weak nuclear force affects leptons (e.g. photons, electrons and neutrinos). If this force were slightly larger, neutrons would decay more quickly and would not be available to form helium.

Helium is necessary for the formation of the heavier elements required for life. If the force were *significantly* larger, hydrogen would burn quickly to helium and only helium would remain to constitute the stars. Without hydrogen, the universe would not contain water, an essential element for life. Similarly, if the weak nuclear force were slightly weaker, no hydrogen would be left. The hydrogen would have become helium and life would not be possible.

The weak force affects the beta decay reaction by which neutrons become protons, electrons and neutrinos. Neutrinos may be the most omnipresent elementary particles in the universe with $\sim 10^9$ neutrinos for each proton and electron. Neutrinos exist in three forms or "flavors" and play an important role in the weak nuclear force. Neutrinos act very weakly with other particles. Neutrinos have no charge and travel at the speed of light. Experimental results place the mass of the neutrino at about 5×10^{-35} kg or about 5×10^{-5} of an electron's mass. Because neutrinos have an enormous density of about $10^9 m^{-3}$ in the universe, the cumulative neutrino mass could exceed the mass of all stars. Accordingly, if there was even a slight increase in the extremely small mass of the neutrino, say 5×10^{-34} instead of 5×10^{-35} kg, the universe would have been a contracting rather than an expanding universe. The almost undetectable mass of the neutrino turns out to be a very finely tuned value.[230]

The galactic material at the universe's beginning is mostly hydrogen and helium. The heavier elements are made inside a star and dispersed when the star ages and explodes as a supernova. This supernova explosion spreads the element rich debris around the galaxy. Carbon, iron, uranium and other heavier elements are the remainder of supernovae. These explosions depend on a very precise value in the weak nuclear force. If the weak nuclear force were much smaller, neutrinos would escape during a supernova explosion and not interact with a star's outer layers. This would preclude the expulsion of the heavier elements necessary for life's formation. Paul Davies concludes:

> If the weak interaction were much weaker, the neutrinos would not be able to exert enough pressure on the outer envelope of the star to cause the supernova explosion. On the other hand, if it were much stron-

ger, the neutrinos would be trapped inside the core, and rendered impotent. Either way, the chemical organization of the universe would be very different.[231]

5.2.9. Precision in the number of dimensions

As discussed above, string theory requires spacetime with at least ten dimensions with only four dimensions (three spatial dimensions and one time dimension) presently flattened out and the remaining six or more curled up or compacted. The laws of physics and chemistry, however, are only compatible with the emergence of life in no more than three spatial dimensions. For example, Stephen Hawking has emphasized that the sun would either collapse into a black hole or disintegrate in any spatial structure exceeding three dimensions. On the other hand, any less than three dimensions would not allow for the emergence of complicated beings such as humans. The digestion process alone would not be possible in a two-dimensional world.[232] John Wheeler agrees and stresses that only a space of three dimensions is complicated enough for the fundamental reactions of life and yet simple enough to avoid the disintegration of life from the effects of quantum physics.[233]

5.2.10. Fine tuning in masses of particles, fundamental values and existence of unchanging types of particles required for DNA

Atomic particles come in stable and unchanging types which allow a DNA molecule to convey information equal to ten thousand pages.[234] Wolfgang Pauli's exclusion principle keeps atomic particles of the same type away from each other and prevents the collapse of atoms. The DNA molecule has a structure which persists because of the balance in the masses of particles and the precision in the values of the fundamental forces. John Leslie discusses the Pauli exclusion principle as an example of how the laws of physics are finely tuned to allow for life:

> The Pauli principle's "spreading out" of the atom by keeping electrons in a fixed hierarchy of orbits is decidedly fortunate. Could electrons take just any orbit

then, (i) thermal buffetings would at once knock them into new orbits, so destroying the fixed properties which underlie the genetic code and the happy fact that atoms of different kinds behave very differently; and (ii) atoms would quickly collapse, their electrons spiralling inwards while radiating violently.[235]

5.2.11. Precision in the agreement between abstract mathematics and the laws of the physical world

Physicist Paul Davies asks why the physical laws of the universe are mathematical. Despite the implications of Gödel's Incompleteness Therorem, he demands an explanation why mathematics work so effectively in applications to the physical world. Mathematics is a language, *une langue bien faite*,[236] which fits extraordinarily well with the physical world. Einstein proposed the general theory of relativity from a strictly abstract mathematical exercise years before it was demonstrated actually to work in the physical world. The agreement between the theory of general relativity and the physical world has been confirmed to more than a trillionth percent precision.[237] Precision to this remarkable degree between counter intuitive abstract mathematical reasoning and the physical world cannot be explained by chance alone.

Consider the agreement between abstract mathematical deductions and the physical world so well described in Pearcey and Thaxton's explanation of Einstein's theory of time dilation. Beginning with James Clerk Maxwell's equations, which predicted that in a vacuum the velocity of light (2.998×10^8 m/s) would remain the same regardless of the velocity of the source of the light, they note that these equations contradict the law of the addition of velocities ($V = v^1 + v^2$) so that an observer watching a train from outside the train would see the light from the train's headlight not moving at a speed of 2.998×10^8 m/s plus the speed of the train, but only at 2.998×10^8 m/s. The velocity of the light is not influenced by the velocity of the train.

The formula for the calculation of velocity ($v = d/t$), where v = velocity, d = distance covered, t = time, does not appear to be accurate when calculating the velocity of light which is a constant. When the train is moving and the engineer switches on the train's headlight, the person watching the train from outside the train will see the light traveling at 2.998 × 10^8 m/s. When the train is moving, the outside observer will see the light covering a greater distance within the same length of time; under the formula "t" is the same but "d" is greater. Normally this would increase the "v." But Maxwell's equations state that the speed of light is not affected by the speed of the source of the light. The only option available mathematically to balance the equation is to change time. Time in the moving train must be slower than time for the person watching the train from outside. This is exactly what Einstein concluded. His theory of time dilation is purely an abstract logical mathematical deduction. If the velocity of light is always 2.998 × 10^8 m/s, always a constant at that speed, when distance changes, mathematically, time must change.[238] Although contrary to intuition, these conclusions have been verified in physical world experiments.

Davies dismisses the argument that the brain imposes mathematical order which does not actually exist in the real physical world; abstract mathematics are too accurate in practical physical applications not to reflect the real nature of the physical world. In referring to the general agreement in this area among his colleagues, he writes, "the belief, which I have found to be held by most scientists is that major advances in mathematical physics really do represent discoveries of some genuine aspect of reality, and not just the organization of data in a form more suitable for human intellectual digestion."[239]

It is difficult to see how natural selection caused the brain to evolve so that it could perform abstract mathematical functions reflecting the real structure of the physical world. Abstract mathematics, like musical ability, has little survival value. Paul Davies recently wrote about the "unreasonable effectiveness" of mathematics in physical science:

> No feature of this uncanny "tuning" of the human mind to the workings of nature is more striking than mathematics. Mathematics is the product of the higher human intellect, yet it finds ready application to the most basic processes of nature, such as subatomic particle physics. The fact that "mathematics works" when

applied to the physical world—and works so stunningly well—demands explanation, for it is not clear we have any absolute right to expect that the world should be well described by our mathematics . . . If mathematical ability has evolved by accident rather than in response to environmental pressures, then it is a truly astonishing coincidence that mathematics finds such ready application to the physical universe. If, on the other hand, mathematical ability does have some obscure survival value and has evolved by natural selection, we are still faced with the mystery of why the laws of nature are mathematical. After all, surviving "in the jungle" does not require knowledge of the *laws* of nature, only of their manifestations.[240]

Before Einstein performed his calculations, the observed universe was explained by Newtonian physics with its Euclidean geometry and mysterious gravitational force. These concepts were derived from man's observance of the world around him. Thus, the image of Newton discovering the law of gravity by watching an apple fall to earth seemed perfectly reasonable from our observance of the action of the gravitational force on falling objects. When the apple left the branch of the tree, one reasoned that it fell to the earth because of the earth's gravitational attractive force. Space was flat, and the curved orbits of the planets moved under the influence of the attractive force of gravity. There was nothing useful for survival which required any radically different thought processes. Newtonian physics worked in the world we experienced.

Einstein's abstract mathematical calculations, however, presented a counter intuitive perspective and exposed a curved spacetime for the universe. The equations of Einstein's theory of general relativity give the following relationship between the density of matter and energy and the curvature of space:

$$8\pi G/C^4 \text{ (density of matter or energy)} = \text{curvature of spacetime}$$

where G is the gravitational constant and C is the speed of light. If there is no matter or energy, and, consequently, the left side of the equation equals zero, space is flat without curvature. The presence of matter causes curvature of spacetime, and matter will move according to that curvature. Where matter is extremely dense, such as in a black hole or in the singularity of the Big Bang, spacetime has a corresponding high degree of curvature.

Curved space and gravity are one and the same in Einstein's equations. The curvature of spacetime influences the motion of matter, and matter influences the curvature of spacetime.[241] Edward Harrison uses the illustration of a stretched, flexible rubber sheet to illustrate the deformation of space. When the rubber sheet is flat, space is flat, and gravity is absent. If a heavy steel ball is placed on the sheet, the rubber sheet becomes curved. Near the ball the curvature is more pronounced with a consequent stronger gravitational force than the gravitational force farther away from the ball where the sheet is more flat. At the distance where the sheet is flat, the gravitational force would be zero. There is no attractive or magnetic force emanating from the ball itself. The gravitational force is the result of the curvature of the rubber sheet caused by the ball's mass. If one were to roll a small steel ball bearing on a completely flat portion of the rubber sheet (assuming that there was no mass present to prevent a completely flat portion), the ball bearing would not move toward the ball; gravity is not a mysterious magnetic like force inherent in the ball which could draw the ball bearing towards the large steel ball. If a ball bearing were rolled on the curved portion of the rubber sheet, however, the ball bearing would assume a curved path around the ball. This curved path would be caused by the curvature of the sheet (which in turn was caused by the mass of the large steel ball), however, not by any inherent attractive force emanating from the ball. (See Figure 5).

The planets and galaxies move in their trajectories according to general relativity's concept of curved spacetime. As John Wheeler has said, "matter tells space how to curve, and curved space tells matter how to move."[242] When spacetime is flat and bodies move at speeds much slower than the speed of light, general relativity and Newtonian physics produce basically the same results. This is true in our solar system where the planets move at relatively low velocities.

The abstract mathematics of general relativity have been demonstrated to be remarkably accurate when applied to the physical world. Einstein's equations meant that gravity was not an attractive force, but was the consequence of the effect of mass on spacetime. A large mass causes space to curve and time to slow down, depending on the observer's distance from the mass. Gravity slows time. According to the equations of general relativity, time should appear to run slower the nearer to a massive body (like the earth) one measures time. Einstein's mathematical deduction has been demonstrated to match perfectly the physical world in a test of relativistic gravitational red shift in the spectrum of light which was measured by comparing the continuous microwave signals generated from space-borne hydrogen maser atomic clocks located in a spacecraft and a similar clock at an earth station. NASA and the Smithsonian Astrophysical Observatory reported that these extremely accurate measurements agreed with the abstract mathematical predictions made according to the general theory of relativity to the 70×10^{-6} level.[243] This is a very precise number. Stephen Hawking noted the incredible accuracy of the prediction from Einstein's equations in another earlier experiment:

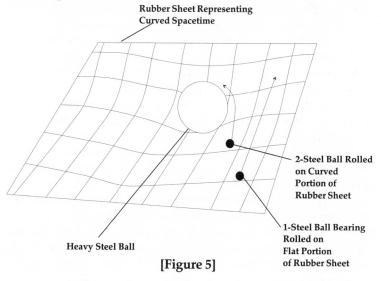

Rubber Sheet Representing Curved Spacetime

2-Steel Ball Rolled on Curved Portion of Rubber Sheet

1-Steel Ball Bearing Rolled on Flat Portion of Rubber Sheet

Heavy Steel Ball

[Figure 5]

> This prediction was tested in 1962, using a pair of very accurate clocks mounted at the top and at the bottom of a water tower. The clock at the bottom, which was nearer the earth, was found to run slower, in exact agreement with general relativity. The difference in the speed of clocks at different heights above the earth is now of considerable practical importance, with the advent of very accurate navigation systems based on signals from satellites. If one ignored the predictions of general relativity, the position that one calculated would be wrong by several miles.[244]

Again, the exceptional performance of abstract mathematical functions in reflecting the real structure of the physical universe appears to have very little survival value. Yet the abstract mathematical functions are verified with astounding accuracy in the physical world. Davies cites Roger Penrose's amazement at the success of abstract, mathematical concepts in describing the physical universe and the unlikely scenario that such abstract theories were the products of accidents:

> It is hard for me to believe, as some have tried to maintain, that such superb theories could have arisen merely by some random natural selection of ideas leaving only the good ones as survivors. The good ones are simply much too good to be the survivors of ideas that have arisen in that random way. There must, instead, be some deep underlying reason for the accord between mathematics and physics, i.e., between Plato's world and the physical world.[245]

Particle physicist and Anglican priest, John Polkinghorne, former President of Queen's College at Cambridge University, notes that science is only possible because the physical world is rationally transparent to us. This transparency is even apparent in counter intuitive phenomena which are revealed by abstract mathematics. He does not consider natural selection processes to be adequate to explain the precise matching of abstract mathematics with the workings of general relativity and with the workings of an unseen subatomic world:

. . . our ability to understand the physical world immensely exceeds anything that is required for the relatively banal purpose of survival. Think of the strange counter intuitive subatomic world of quantum theory. If you know where an electron is, you cannot know what it is doing; if you know what it is doing, you cannot know where it is. That is Heisenberg's uncertainty principle in a nutshell. The quantum world is totally unpicturable for us, but it is not totally unintelligible. I cannot believe that our ability to understand its strange character is a curious spin-off from our ancestors having had to dodge saber-toothed tigers. That seems even clearer when we recognize that it is *mathematics* which gives us the key to unlock secrets of nature. Paul Dirac spent his life in the search for beautiful equations. That is a concept not all will find immediately accessible, but among those of us who speak the language of mathematics, mathematical beauty is a recognizable quality. . . . Time and again we have found that it is equations with that indispensable character of mathematical beauty which describe the nature of the physical world. If you stop to think about it, that is a very significant thing to have discovered. After all, mathematics arises from the free rational exploration of the human mind. Yet it seems that our minds are so finely tuned to the structure of the universe that they are capable of penetrating its deepest secrets.[246]

5.2.12. Conclusion: *abundance of evidence from precision of values against accidental formation of a universe compossible with life*

We can imagine many universes similar to our own which would contain very slightly different force strengths or values which would preclude the formation of life. These strengths and values must be present in combinations and permutations within very low mathematical probability constraints. We have discussed many examples, but the foregoing is only a partial list of some of the aspects of fine tuning in our universe necessary for the development of life. As noted many times, the mathematical probabilities against these precise conditions occurring by accident or chance alone are overwhelming. Each year new discoveries in physics bring more evidence against the accidental formation of a universe consistent with the formation of life. John Leslie summarizes the wealth of evidence against accident contained in the immense precision of values and laws required for the emergence of life:

> Those seeking evidence of fine tuning may appear to have embarrassingly much at which they can point. A force strength or a particle mass often seems to need to be more or less exactly what it is *not just for one reason, but for two or three or five*. Yet obviously, it could not be tuned in first one way and then another, to satisfy several conflicting requirements. A force strength or a mass cannot take several different values at once! . . . One possible response would be that when factor *A* looks as if it needed fine tuning in order to bring it into life-generating harmony first with factor *B*, then with factor *C*, and then with *D*, and so on, what really occurred was the reverse. It was factors *B*, *C*, and *D* which were all of them fine tuned so as to harmonize with *A*. My hunch is that while such a response has some force, it is not by itself enough. After responding in that way in the case of factor *A*, we would find ourselves under pressure to say the same kind of thing about factor *B* as well—but that would lead to inconsistency.[247]

5.3. Speculations to avoid a beginning out of true nothingness

Many scientists do not like the metaphysical implications of a beginning to the universe or some of the probability constraints that a beginning requires. Setting aside the rule of Ockham's razor, these scientists are attracted to several speculations which would avoid a beginning out of nothing and appeal to infinity or other concepts which could increase the mathematical probabilities for accidental processes in the formation of the universe. For example, the concept of an oscillating universe would allow for an infinite number of beginnings.

5.3.1. Conjecture of an oscillating universe

An oscillating universe would be a closed universe with an endless cycle of contractions and expansions. This would mean that the universe had no beginning and consequently did not need a cause and that the universe could avoid a singularity and survive the Big Crunch. The Second Law of Thermodynamics, however, would require that each oscillating universe would become more disordered and chaotic. After several cycles, the disorder and entropy would preclude another expansion. The possibility of an oscillating or bouncing universe was removed by Roger Penrose and Stephen Hawking, who demonstrated that the gravitational force on a collapsing universe would produce a singularity which would not allow for another expansion.[248]

In his book, *The Emperor's New Mind*, Oxford Professor Penrose gives an interesting analysis of the uniqueness of the Big Bang singularity. As indicated above, for the sake of performing the calculations, one makes the assumption that the universe is closed and a Big Crunch is inevitable. The entropy or disorder at the Big Crunch can be computed by following the formula for the entropy at the singularity of a black hole. We will now discuss Penrose's computations as they relate to the concept of an oscillating universe.

Penrose notes that Sir Arthur Eddington calculated the number of baryons (protons and neutrons) in the observable universe at 10^{80}. This is certainly a conservative estimate and if the number were larger the entropy or disorder at the singularity would be even greater. Using the estimate of 10^{80}, Penrose uses Jacob Bekenstein and Stephen Hawking's formula for calculating the entropy of a black hole.[249] According to this formula, the entropy of a black hole is related to the surface area of its horizon. When a black hole is spherically symmetrical, the surface area is related to the square of the mass of the black hole.[250] The entropy of the black hole is proportional to the square of its mass so that the entropy increases with the size of the black hole. The greatest entropy occurs when the most mass (or energy since $E=mc^2$) has collapsed into the black hole.

If one assumes that the universe is closed and will consequently be drawn back into a Big Crunch, we can use the Bekenstein-Hawking formula to estimate the entropy at the Big Crunch. If we assume that Eddington's estimate of the number of baryons (protons and neutrons) in the universe is correct at 10^{80}, Roger Penrose's calculation of the entropy per baryon is 10^{43} for an incredible entropy at the Big Crunch of 10^{123}. If Eddington's figure were larger, under the Bekenstein-Hawking formula, the entropy would also be larger. Thus, the Big Crunch would be totally chaotic. This disorder would preclude another expansion.[251]

5.3.2. Conjectures of quantum fluctuations, wave functions and no boundaries

5.3.2.1. The curtain at Planck time.   Prior to some empirical confirmation of the Big Bang theory, the dominant theory of the universe was the steady state theory which held that the universe had always existed and time plus chance could allow for almost anything. As discussed above, however, the calculations by George Gamow, results from the Cosmic Background Explorer concerning the uniformity of microwave background radiation ripples, and even Einstein's admission of his blunder in using the cosmological fudge factor, destroyed the steady state theory. The confirmation of the Big Bang shows that the universe had a beginning and that it did not always exist. We have already discussed the impossibility of an oscillating universe alternating between cycles of Big Bang and Big Crunch. So where did the universe come from? Matter, space and time came into existence with the Big Bang. What existed "before" the bang?

We cannot move our observations or calculations of the universe backwards in time beyond Planck time or 10^{-43} of the first second after the Big Bang. Before Planck time the universe would still be smaller than a proton, the temperature would be 10^{32} degrees K, and the particles of quantum physics could not have existed. (As indicated above, technically, we cannot even use the terms "before", "prior" or "pre" in relation to Planck time, because time begins at Planck time which is time zero.)

The laws of physics and our understanding of quantum particles limit our ability to speculate as we approach the singularity of time zero of the Big Bang. At Planck time or time zero the theories of physics fail completely. Quantum physics and other classical theories of physics no longer suffice to describe the state of the universe.[252] The dimensions of the universe decrease as we extrapolate back in time towards the singularity. As the size of the universe decreases to less than the size of a proton, the quantum fluctuations of gravity become so large that the classical laws of physics can no longer be valid. This problem occurs when the dimension of the universe is 1.6×10^{-33} centimeters. Assuming that the universe expanded from an infinite density, this dimension would be reached at Planck time with the density of the universe at 10^{94} grams per cubic centimeter. We can only speculate about the state of the universe "prior" to this Planck time which, as noted above, is time zero when the cosmic clock began to tick.[253] As John D. Barrow has written recently:

> The Planck time of 10^{-43} seconds is significant, because when we reach this extraordinarily early time the size of the visible universe becomes smaller than its quantum wavelength and is thus enshrouded by quantum uncertainty. When quantum uncertainty overtakes everything, we don't know the positions of anything, and we can't even determine the geometry of space. This is when Einstein's theory of gravitation breaks down.[254]

Something happened before Planck time, but the equations of science cannot speak to it. Robert Jastrow, an agnostic astronomer, has concluded that the Big Bang began under such conditions that appear to make it impossible now or ever to determine what force or forces brought the universe into being. In his book, *God and the Astronomers*, Jastrow concluded:

> For the scientist who has lived by his faith in the power of reason, the story ends like a bad dream. He has

scaled the mountains of ignorance; he is about to con-
quer the highest peak; as he pulls himself over the final rock, he is greeted by a band of theologians who
have been sitting there for centuries.[255]

All the laws of classical physics, including the law of general
relativity, without a quantum theory of gravity, break down at
Planck time.[256] All scientific theories about "pre" Planck time are
purely speculative at present.

5.3.2.2. Quantum fluctuations. One speculation is that the universe began as a quantum fluctuation out of nothing. Joseph Silk
defines the "nothing" as a quantum vacuum. Pursuant to Heisenberg's principle of uncertainty, as it operates in post Planck time
quantum physics, a vacuum is not really empty and contains
matter. Because the particles of this matter are not directly observable, they are called "virtual" particles. These virtual particles
are continuously appearing and disappearing in very short
bursts of time. This follows from Einstein's well known $E=mc^2$
formula which means that energy can create mass and vice versa.
In the quantum world energy or mass appear and then quickly
disappear. Under the Heisenberg principle a quantum particle
can appear briefly, provided it is quickly converted back into energy. The post Planck time universe is filled with these short-
lived particles (and anti-particles) continuously being created out
of what appears to be nothing and then disappearing in a very
small fraction of a second. Material particles are created out of
quantum energy fluctuations. The mathematics behind the quantum fluctuation theory are relatively simple. Because the speed of
light is a constant, any uncertainty in energy must be the result of
an uncertainty in mass. Thus,

$$\Delta E = c^2 \Delta m$$

Heisenberg's principle for the uncertainty in the momentum
or position of an electron has an analogy to the uncertainty of the
precise energy of a quantum physical system at any given moment in time. Consequently, one cannot know the amounts of
energy in the quantum world over short intervals of time. Thus,

$$\Delta E \times \Delta t \geq h/2\pi$$

Applying this equation to the uncertainty in mass, we have,

$$\Delta m \times \Delta t \geq h/2\pi c^2$$

This equation means that in an extremely short period of time, the amount of matter in a particular amount of space is uncertain. Matter can spontaneously appear and disappear during this very brief time. This is what actually happens in space; particles and antiparticles spontaneously appear and then disappear. To calculate the duration of time that an antiparticle and a particle can exist, William J. Kaufmann III assumed that an electron and an antielectron each had a mass of 9.1×10^{-31} kg. Following the above equation and substituting the combined mass of the particles for Δm, Δt would equal 6.5×10^{-22}s. This would be the maximum interval of time during which the antiparticle and particle could appear without violating the principle of the conservation of mass. As the size of these particles increases, the period of time in which they can exist shortens proportionally. The mass of a proton is 1,836 times the mass of an electron so a proton and an antiproton can only appear for 1/1,836 as long as their electron/antielectron counterparts.[257]

In other words, the "virtual" particle would borrow energy from its source for a fleeting duration. The shorter the duration, the larger the potential "virtual" particle. Since the amount of time for the duration of these particles varies inversely with their mass, the duration for a system with the mass of the universe would be well short of 10^{-43}s, not the billions of years required by estimates of Hubble's constant.

In 1973 physicist Edward Tryon, however, proposed that if the total mass-energy content of the universe was equal to zero, a quantum fluctuation could occur which would allow for a particle to exist for an indefinite time.[258] Tryon based his reasoning on calculations which indicate that a star's total energy content is zero with the content of the negative gravitational energy matching the content of the star's matter-energy. If the total energy-mass content of a star (or of the universe) is zero, the First Law of Thermodynamics (the law of conservation of mass and energy:

energy and matter can neither be created nor destroyed) would not be violated if a star (or the universe) came into existence by a quantum fluctuation from a quantum field.

Although Tryon's proposal appears plausible, the probability of the appearance of the universe from a quantum fluctuation is extremely remote and perhaps zero. This zero probability is based on the equation of quantum mechanics which relates a finite time period with a finite probability. As Hugh Ross argues:

> Quantum mechanics is founded on the concept that quantum events occur according to finite probabilities within finite intervals of time. The larger the time interval, the greater the probability that a specific quantum event will occur. This means that if the time interval is zero, the probability for that quantum event occurring is also zero. Because time began when the universe was created, the time interval is zero, eliminating quantum tunneling as a possible candidate to be the creator of the cosmos.[259]

The process of taking an occurrence from the quantum microworld, applying it to the entire cosmos, and speculating that the entire universe came into being by a quantum energy fluctuation presents more questions than it answers. Similar to the paragraph quoted in the reasoning section above from Mark Twain's *Life on the Mississippi*, the extrapolation from the appearance of virtual particles in space raises a multiplicity of issues. Why would a quantum energy fluctuation occur when the laws of quantum mechanics have broken down?

As the above quotation from John D. Barrow indicates, we don't even know the geometry of space when the universe is less than its quantum wavelength. At that size all events are enshrouded in uncertainty. One must then ask the question why should the Heisenberg principle, based on an analysis of conventional time and space (which did not exist until Planck time), function in some unknown way to cause the formation of the universe under nonconventional timespace conditions about which we have no knowledge? Certainly this is speculation squared if not cubed. Many, if not most, physicists do not take the concept seriously:

> Of course, some will argue that since we do not know exactly what occurred before the universe was 10^{-43} second old, the possibility necessarily exists that the relationship between time and the probability for certain quantum events in that tiny time interval could break down. However, this argument is based on pure speculation, actually multiple speculations. First one must speculate that the breakdown occurred at precisely the needed moment of time and location of space. Finally, one must speculate that this breakdown occurred in such a fashion that the quantum tunneling of the entire universe took place.[260]

Even assuming the validity of the speculation on the formation of the universe from something analogous to a quantum energy fluctuation pursuant to the laws of quantum physics, where did these laws come from? John Polkinghorne notes: "A quantum vacuum is not *nihil*. Where do quantum mechanics and the fluctuating fields including those of general relativity (the generalization of space) come from?"[261] For the purposes of the first question presented, the issue whether the universe had a beginning in a singularity or appeared as the result of something analogous to a quantum fluctuation, is not dispositive of the question. *Ex nihilo nihil fit*.[262] As Oxford Professor Keith Ward writes: "On the quantum fluctuation hypothesis, the universe will only come into being if there exists an exactly balanced array of fundamental forces, an exactly specified probability of particular fluctuations occurring in this array, and an existent space-time in which fluctuations can occur. This is a very complex and finely tuned nothing !"[263]

Quantum cosmologists have much to explain: quantum fluctuations need a context of space and time, a perfectly balanced zero net energy from a matching negative gravitational energy and a positive kinetic and rest mass energy, a quantum field with certain characteristics of mass and energy, and the laws of quantum mechanics which dictate precise probabilities to fluctuations in this background.[264] M.A. Corey's analysis is consistent with Ward's position as he presents a theist's perspective on a quantum fluctuation as the basis of cosmogenesis:

> This view (non-theistic cosmogenesis) is fallacious, however, because sudden quantum appearances don't really take place out of "nothing." A larger quantum field is first required before this can happen, but a quantum field can hardly be described as being "nothing." Rather, it is a thing of unsearchable order and complexity, whose origin we can't even begin to explain. Thus, trying to account for the appearance of the universe as a sudden quantum fluctuation doesn't do away with the need for a Creator at all; it simply moves the whole problem backwards one step to the unknown origin of the quantum field itself.[265]

Even if one assumes the validity of the quantum fluctuation hypothesis, the formation of a universe compossible with life could not result from accident alone. Chance alone was not sufficient; the laws of physics were necessary. As Heinz Pagels has noted:

> The nothingness "before" the creation of the universe is the most complete void that we can imagine—no space, time or matter existed. It is a world without place, without duration or eternity, without number—it is what the mathematicians call "the empty set." Yet this unthinkable void converts itself into the plenum of existence—a necessary consequence of physical laws. Where are these laws written into that void? What "tells" the void that it is pregnant with a possible universe? It would seem that even the void is subject to a law, a logic that exists prior to space and time.[266]

Who or what designed the laws of physics? Albert Einstein wrote that the natural law "reveals an intelligence of such superiority that, compared with it, all the systematic thinking and acting of human beings is an utterly insignificant reflection."[267] Stephen Hawking asked the questions: Where do the laws of physics come from? What is it that breathed the fire into the equations? As physicist Freeman Dyson commented on the fine tuning of the universe: "The more I examine the universe and the details of its architecture, the more evidence I find that the universe in some sense must have known we were coming."[268]

Where do the laws of physics (and the resulting laws of chemistry) come from? If we say that they have always existed, we know that the laws as presently understood break down at Planck time. If string theory is true, then we have the issue of 10 or, if the cause of the universe is outside the dimensions of the universe, 11 dimensions. Most of the laws required for life only work in four dimensions. If the ten dimensions split into four and six dimensions at Planck time, then the laws as we know them could not have existed in the period when there were 10 dimensions. What was the logic or intelligence that existed prior to space and time? *Vere scire est per causas scire.*[269]

5.3.2.3. Wave function and the no boundary proposal. In 1970 Roger Penrose and Stephen Hawking published a paper which proved that in any expanding universe where the theory of general relativity applied and the universe contained as much matter as we observe, a Big Bang singularity must have existed. The beginning of time was a point of infinite density and infinite curvature.[270] In 1984 Hawking and James Hartle authored a paper which became the basis for Hawking's no boundary proposal (which he emphatically stresses is "just a proposal") to the effect that a singularity might not exist under certain questionable presuppositions in "imaginary" time, a defined mathematical concept. According to this proposal, a singularity could be avoided by a quantum mechanical wave function provided certain assumptions are made which are contrary to our understanding of the quantum state of the universe.

In their joint paper, Hawking and Hartle attempted to apply a unique and novel application of quantum physics to the universe as a whole. Rather than applying quantum mechanics to quantum particles, they proposed applying the principles of quantum mechanics to the creation of space and time. To avoid a singularity at the beginning of time, they used the analogy of a

hydrogen atom examined from the perspective of a quantum mechanical wave function. The singularity is avoided when the hydrogen atom is described by a probability wave function.

A quantum equation developed by Erwin Schrödinger shows
that the probability of an electron's location depends upon the
sum of the trajectories which are determined by the magnitude
and the phase of the waves that distinguish those trajectories.[271]
Schrödinger's probability wave function applies to all matter. For
large objects the wave function is not significant and the location
of a car or a rocket ship is not reduced to a calculation of probabilities. On a much smaller level, however, such as a subatomic level
or in the smallness of the initial compression of the universe to a
point smaller than an atomic particle, probability calculations are
required.

Hawking and Hartle proposed calculating the wave function
for the whole universe as one would calculate an electron's wave
function. They speculated that when the universe was in a state
of minimum excitation (vacuum state), the singularity could disappear just as a wave function description avoids a singularity in
an atom with one electron and one proton (the hydrogen atom).
Caret initio et fine.[272] In their own words:

> In the classical theory the singularity is a place where
> the field equations, and hence predictability, break
> down. The situation is improved in the quantum the-
> ory. An analogous improvement occurs in the prob-
> lem of an electron orbiting a proton. In the classical
> theory there is a singularity and a breakdown of pre-
> dictability when the electron is at the same position as
> the proton. However, in the quantum theory there is
> no singularity or breakdown. In an *s*-wave state, the
> amplitude for the electron to coincide with the proton
> is finite and non-zero, but the electron just carries on
> to the other side. . . . The ground-state wave function
> in the simple mini-superspace model that we have
> considered with a conformally invariant field does
> not correspond to the quantum state of the universe
> that we live in because the matter wave function does
> not oscillate. However, it seems that this may be a
> consequence of using only zero mass fields and that

the ground-state wave function for a Universe with a massive scalar field would be much more complicated and might provide a model of quantum state of the observed Universe. If this were the case, one would have solved the problem of the initial boundary conditions of the Universe: the boundary conditions are that it has no boundary."[273]

In his subsequent book, *A Brief History of Time*, Hawking is very cautious in describing his speculations which attempt to avoid a singularity and a beginning to the universe. He admits that his proposal is *plus in posse quam in actu*.[274] In his book he attempts to use imaginary numbers to circumvent a beginning. Hawking uses imaginary time as the dimension for his calculations involving imaginary numbers which, of course, are a valid and useful mathematical concept.[275] This controversial approach appears to many physicists as a mathematical contrivance or trick to arrive at a conclusion consistent with his metaphysical predilection for avoiding a singularity. The universe he describes exists only in mathematical terms and apart from real spacetime. He emphasizes the speculative nature of his concept and stresses that it only functions in imaginary time and not in real time:

> I'd like to emphasize that this idea that time and space should be finite without a boundary is just a *proposal: it cannot be deduced from some other principle.* Like any other scientific theory, it may initially be put forward for aesthetic or metaphysical reasons, but the real test is whether it makes predictions that agree with observation . . . If the universe really is in such a quantum state, there would be no singularities in the history of the universe in imaginary time. It might seem therefore that my more recent work had completely undone the results of my earlier work on singularities. *But . . . when one goes back to the real time in which we live, however, there will still appear to be singularities.* The poor astronaut who falls into a black hole will still come to a sticky end; *only if he lived in imaginary time would he encounter no singularities . . . In real time, the universe has a beginning and an end at singulari-*

ties that form a boundary to space-time and at which the
laws of science break down.[276] (emphasis added)

Hawking appears to be using a kind of regressive reasoning which we discussed in our section on logic. Although a valid approach to assist in creating plausible hypotheses, this kind of reasoning requires some verification. As we discussed earlier, reasoning backwards is useful, but it must be verified by an antecedent which is *secundum veritatem.*[277] Hawking indicates that his proposal may have been put forward to coincide with metaphysical predilections and admits that his proposal cannot even be deduced from any verified principle. Nevertheless, he speculates that imaginary time may be more fundamental for an understanding of the universe than ordinary time. He proposes that time really was like space in the very early universe. At an instant in imaginary time, space and time dimensions were identical. Oxford Professor Keith Ward is unimpressed with Hawking's use of imaginary time:

> According to the Hartle/Hawking model . . . time itself is signified by a complex number (part of which involves an imaginary number, such as the square root of a negative number), and it becomes an internal property of a set of three-spaces. I do not think this can any longer rightly be called "time" at all, in any sense we can recognize it. What has happened is that the phenomenological reality of time has been transformed into a mathematical variable, and then treated as a pure abstraction, which, far, far from giving the "true reality" of time, has less and less relation to the real time one started from. The conceptual problems of such a model are enormous . . . It is only because he does not ask why the quantum laws are as they are that he can say that the universe is not affected by anything outside its own parameters . . . The physical existence of this universe, even on highly disputable quantum gravity theories such as those of Hawking, is due either to extraordinary chance or to a choice from possible mathematical structures of extraordinary precision.[278]

Kitty Ferguson points out that Hawking's no boundary proposal actually has boundaries. She notes the conventional definition of boundary conditions which comprise the initial conditions in an experiment. These initial conditions may not have boundaries in space and time, but are still boundary conditions in the "underlying context of logic and laws, the specification required in order for the proposed situation to exist at all."[279] In other words, the universe proposed by Hawking, which would have no boundaries in space and time, could only exist if Hawking proposed boundary conditions of underlying logic which would be required for the existence of such a universe. Ferguson has her doubts about his proposal:

> Hawking is the first to point out that his idea is just a proposal. He doesn't even call it a theory. It's a spectacularly wild leap of imagination. He hasn't deduced these boundary conditions from some other principle. . . . mathematical and logical consistency do not demand this model of the universe as opposed to others. Nothing has so far shown that it is the only consistent model or one to be strongly preferred over others. Could it have happened this way? It's far too early in the game to answer that question. Did it happen this way? Only on aesthetic and philosophical grounds, and because it upholds one of the assumptions of science, is it possible at present to prefer this theory over others.[280]

5.3.2.4. Hawking's question and the need of a creator as causa essendi. Not everyone is satisfied that Hartle and Hawking's no boundary proposal is internally consistent. For readers with theistic concerns we should note that Don Page, one of Hawking's collaborators, has used the example of an artist's drawing of a circle to illustrate that the absence of a beginning or an end does not remove the artist as the cause of the circle.[281] The issue of a beginning to the universe is not necessarily fundamental to the question of the existence of God. Readers with religious faith who believe in the reality of God need not view Stephen Hawking's no boundary proposal as a battle ground. In referring to the absence of a singularity or a beginning in his proposal,

Hawking asks the rhetorical question, "What place, then, for a creator?" The answer I assume Hawking is making to this rhetorical question fails to distinguish between *causa essendi* (a cause of existence) and *causa fieri* (a cause of becoming). Something which exists may need a cause for its continuing existence without necessarily needing a cause for its becoming. Even assuming, *argumenti causa*,[282] that the no boundary proposal reflects reality, a creator who is a necessary and non contingent being is required as a *causa essendi* for the continued existence of the universe pursuant to the following reasoning:

1. To avoid the fallacy of *petitio principii*, assume that the universe exists without a beginning. (If we assume a beginning, we beg the question of a creative cause). This is consistent with Hawking's proposal and is perhaps the main motivation behind it.
2. A distinction must be made between *causa essendi* and *causa fieri*. A mare may be *causa fieri* of her foal, but a mare does not act as *causa essendi* of her foal; she is not the cause of the continuing existence of her foal. A mare which passes away while her foal continues to inhabit the earth cannot be the cause of the foal's continuing existence. A match may be *causa fieri* of a flame, but oxygen acts as *causa essendi* because it is a necessary condition for the continuing existence of the flame.
3. Something which needs a cause of its continuing existence at every moment is *contingent* upon that cause; it is not *necessary* in and through itself.
4. As we have discussed, this universe is only one among many possible universes which might have existed. We can conceive of other universes which could exist with different characteristics than our universe. Because other universes are possible, this universe is not the only universe that could ever exist. It is not a *necessary* universe. Because it is merely a *possible* universe and not a necessary universe, its existence is not necessary in and through itself. It is not the only universe which can ever exist.
5. Something must exist when it cannot be anything except what it is; it cannot not exist. It is *necessary*. However, some-

thing which could be other than what it is might not exist. A universe which could be other than what it is might not be at all. Such a universe has the possibility or the potential for non-existence.

6. A universe which has the potential for non-existence is a *contingent* rather than a *necessary* universe. Anything that is contingent requires a *causa essendi*, an effective cause of its continuing existence. This merely possible universe is contingent and requires a *causa essendi* to prevent the possibility of its non-existence. This merely possible universe requires a preservative cause of its continuing existence to protect it from the possibility of annihilation (its reduction to nothingness). This preservative activity is an action of ex-nihilation (coming out of existence out of nothing) as it is juxtaposed to an action of annihilation.

7. Even if Hawking's boundary-less proposal is correct (which is very unlikely) and the universe does not need a *causa fieri* for its coming into existence, it does need a *causa essendi* for its preservation and to protect it from the possibility or potential of a reduction to nothingness or annihilation.

8. To prevent annihilation, the *causa essendi* cannot be a natural cause because natural causes are themselves contingent things. Contingent things cannot act as *causa essendi* because they do not have the cause of their own continuing existence in themselves. Something that is *necessary and uncaused* is required to act as *causa essendi* of a *contingent* thing.

9. If we define the concept of God as a necessary rather than a contingent being, God cannot be part of the universe, because the universe and all of the individual things in it are contingent in their existence. A necessary existence means that such an existence is uncaused, independent and unconditioned. In this concept God has a necessary existence.

10. Thus, even if we assume that Hawking's questionable proposal is true, the answer to his presumed rhetorical question concerning the need for a creator is that a creator is necessary as a preservative cause of the existence of the universe.

The important premise in this argument is that the universe is contingent and not necessary. Because other universes are possible, our universe is not necessary in and through itself. If it is not necessary, it is contingent. As Professor Keith Ward argues:

> To say that the existence of this universe is necessary is to say that no other universe could possibly exist. But how could one know that, without knowing absolutely everything? Even the most confident cosmologists might suspect that there is something they do not know. So it does not look as though the necessity of this universe can be established. . . . The physical cosmos does not seem to be necessary. We can seemingly think of many alternatives to it. There might, for instance, be an inverse cube law instead of an inverse square law, and then things would be very different, but they might still exist. We can see how mathematics can be necessary, but it is a highly dubious assertion that there is only one consistent set of equations which could govern possible physical realities. We cannot bridge the gap between mathematical necessity and physical contingency. How could a temporal and apparently contingent universe come into being by quasi-mathematical necessity?[283]

In his book, *How to Think About God,* Mortimer Adler stated this argument and his position that this premise cannot be affirmed with certitude but only beyond a reasonable doubt. He concluded that a preponderance of reasons favor the belief that God exists. Adler himself was persuaded that God exists either beyond a reasonable doubt or by a preponderance of reasons.[284]
The reasoning for a *causa essendi* for the preservation of the universe is consistent with a God whose involvement in the universe is continuous. As *causa essendi* God would not be simply a cause which began or wound up a universe compossible with life and then left it to run on its own, but a cause which intimately and constantly preserves the universe in all of its detail. With respect to the existence of God, one may argue that each person must make an act of free choice in determining his own conclusions. The reasoning on either side of this choice does not produce an absolutely compelling argument. Either conclusion requires a

leap of faith. It is up to each individual to decide in which direction he or she will leap. No perfectly ironclad argument destroys the freedom to make that decision. Owen Gingerich appears to agree with this assessment in describing a discussion he had with Freeman Dyson and concludes:

> From a Christian perspective, the answer to Hawking's Query is that God is more than the omnipotence who, in some other space-time dimension, decides when to push the mighty ON switch. A few years ago I had the opportunity to discuss these ideas with Freeman Dyson, one of the most thoughtful physicists of our day. "You worry too much about Hawking," he assured me. "And actually it's rather silly to think of God's role in creation as just sitting up there on a platform and pushing the switch." Indeed, creation is a far broader concept than just the moment of the Big Bang. God is the Creator in the much larger sense of designer and intender of the universe, the powerful Creator with a plan and an intention for the existence of the entire universe. The very structures of the universe itself, the rules of its operation, its continued maintenance, these are the more important aspects of creation. Even Hawking has some notion of this, for near the end of his book he asks, "What is it that breathes fire into the equations and makes a universe for them to describe? The usual approach of science of constructing a mathematical model cannot answer the question of why there should be a universe for the model to describe. Why does the universe go to all the bother of existing?"[285] Indeed, this is one of the most profound, and perhaps unanswerable, theological questions.

Theological implications notwithstanding, when describing his no boundary proposal, Hawking emphasizes that when one returns to real time, the universe has a beginning and an end. Roy Peacock, a professor of aerospace science, notes that this conclusion fits well with the application of the Second Law of Ther-

modynamics to the universe as a whole.[286] We will now consider
that application and its meaning for Hawking's no boundary proposal.

*5.3.2.5. The no boundary proposal and the Second Law of Thermo-
dynamics.*   Hawking appears to give only peripheric analysis to
certain issues raised by the Second Law of Thermodynamics in
his symmetrical boundary less proposal. As noted above, the Second Law requires that the entropy or disorder in the universe as
a whole tends toward a maximum. If the universe is moving towards a maximum entropy, then a minimum entropy must have
existed. The minimum was a starting point or a beginning. The
Second Law requires a continuing overall reduction in order as
the universe proceeds to dissipate or run down. A universe with
an overall principle of entropy moving towards a maximum indicates that such a universe must have a starting point of minimum
or lower entropy (maximum or higher order). The Second Law's
direction of increasing disorder can be viewed as an arrow of
time that always moves toward the future. Although other laws
of physics are time reversible, the Second Law is not. Increasing
entropy is one of the physical laws which distinguish past from
future.[287] The Second Law requires a beginning and an end to the
universe. At the end the universe will reach maximum entropy,
and time will cease. In his book, *The Quark and the Jaguar*, Nobel
Prize winning physicist Murray Gell-Mann comments on the
high level of order required by the Second Law at the beginning
of the universe:

> A deeper question is why the same argument is not
> applicable when the direction of time is reversed.
> Why should a film for a system run backwards not
> show it moving toward probable disorder instead of
> toward order? The ultimate answer to that question
> lies in the simple initial condition of the universe at
> the beginning of its expansion some ten billion years
> ago, contrasted with the condition of indifference
> that is applied to the distant future in the probability
> formula of quantum mechanics. It is not only the
> causal arrow of time that points from past to future as

> a result but also the other arrows, including the or-
> der-disorder or "thermodynamic" arrow of time. . . .
> Most large-scale order in the universe arises from or-
> der in the past and ultimately from the initial condi-
> tion. That is why the transition from order to the
> statistically much more probable disorder tends to
> proceed everywhere from past to future and not the
> other way around. We can think of the universe met-
> aphorically as an old-fashioned watch that is fully
> wound at the beginning of its expansion and then
> gradually runs down while spawning smaller, partial-
> ly wound watches that slowly run down in their turn,
> and so on.[288]

If many more disordered states are probable than ordered states, why, as discussed above, did the Big Bang produce a universe with such a high degree of order? Roger Penrose demonstrated that in any Big Crunch entropy would increase so that a Big Crunch would result in total chaos. Hawking initially thought that entropy might decrease in a contraction but changed his mind and admitted that he had made a mistake:

> As I said, I thought at first that the no boundary con-
> dition did indeed imply that disorder would decrease
> in the contracting phase. . . . I realized that I had
> made a mistake: the no boundary condition implied
> that disorder would in fact continue to increase dur-
> ing the contraction. The thermodynamic and psycho-
> logical arrows of time would not reverse when the
> universe begins to recontract or inside black holes.[289]

Hawking's reversal of his position on the question of the state of entropy in any Big Crunch presents a logical dilemma which Huw Price does not believe he has solved. In an article in *Nature* entitled, "A Point on the Arrow of Time," Price argues that Hawking does not explain how his no boundary proposal can be consistent with the concept of a low entropy at the Big Bang without implying a low entropy in a Big Crunch or in a black hole. How does Hawking derive an asymmetrical consequence (low entropy near the Big Bang and high entropy in the Big Crunch) from his symmetrical physical proposal?[290]

Either Hawking's symmetrical no boundary proposal excludes high entropy at both extremes of time (at the Big Bang or at the Big Crunch) or his proposal does not exclude high entropy at both of these extremes. In the latter case, as described in the above quotation, Hawking has already written that high entropy or disorder would increase in the Big Crunch so only the latter option is available. But Hawking has acknowledged the low entropy (high order) state at one end of the thermodynamic arrow of time. Hawking poses the following questions:

> But why should the thermodynamic arrow of time exist at all? Or, in other words, why should the universe be in a state of high order at one end of time, the end that we call the past? Why is it not in a state of complete disorder at all times? After all, this might seem more probable. And why is the direction of time in which disorder increases the same as that in which the universe expands?[291]

Hawking's predicament is that if he does not support a reversal of the thermodynamic arrow of time in the circumstances of a Big Crunch or other massive gravitational collapse, temporal asymmetry (one directional time) cannot be explained because our best physical theories cannot account for the low initial entropy of the universe.[292] He does not explain how his symmetrical no boundary proposal can allow for the asymmetrical implications of low entropy near the Big Bang and high entropy near the Big Crunch. Huw Price explains Hawking's dilemma:

> As I see it, the other possibilities are that Hawking has made one of two mistakes. Either his no-boundary proposal does exclude disorder at both temporal extremities of the Universe, in which case his mistake was to change his mind about contraction leading to decreasing entropy; or the proposal does not exclude disorder at either temporal extremity of the Universe, in which case his mistake is to think that the no-

boundary proposal does away with the need for initial conditions in explaining temporal asymmetry.[293]

In his book, *Time's Arrow and Archimedes Point*, Price relates that Hawking has not answered these issues in a direct manner, but, at a conference on time's arrow, Hawking proposed that entropy decreases when the universe is small and increases when the universe is larger. Hawking refers to the "early" time in the expansion of the universe and to how the universe "starts off" in a particular state. Price demands consistency in Hawking's use of language in describing the no boundary proposal:

> How are we to interpret these references to how the universe *starts off*, or *starts out*, or to the *early* universe? Do they embody an assumption that one temporal extremity of the universe is objectively its start? Presumably Hawking would want to deny that they do so, for otherwise he has simply helped himself to a temporal asymmetry at this crucial stage of the argument . . . But without the assumption that one temporal extremity of the universe is "really" the beginning, what is the objective content of Hawking's conclusion? Surely it can only be that the specified results obtain when the universe is small—as Hawking's own gloss has it . . . in which case the argument must work at both ends of the universe, or at neither . . . The only way to make the argument coherent is to take it to apply to any temporal extremity, but in this case the consequences of the no boundary condition will be symmetric: if one end of the universe has to be ordered, so must the other be.[294]

Whether Hawking has an answer to Price's objections remains to be seen. Nevertheless, the improbability of a universe with a low entropy (highly ordered) beginning argues against accident as the cause of its formation. The Second Law requires a process of overall reduction of order in the universe and is not a secondary law in physics, but one of the highest of the physical laws. Roy Peacock argues that unlike other laws, the Second Law does not depend upon initial conditions or operational constraints. The consequences will always result in an overall increase in entropy. Because it is the only law that defines time's direction, entropy must increase with time. This requires a beginning to the universe. In Peacock's own words: "We have used this law in determining that the universe had a beginning, creation. Even though the tools of the physicist are unable to break into the secrets of the first moment, we can conclude that it initiated a period of low, but increasing entropy."[295]

5.4 Weak and strong anthropic principles

The weak anthropic principle basically states that the coincidences, balances, and fine tuning in particle astrophysics must be the way they are or we would not be here to think about or observe them. The principle, however, is not an explanation for the precise values seen everywhere in the universe. As Joseph Silk, Professor of Physics at the University of California at Berkeley admits:

> Indeed, some cosmologists think that such an anthropomorphic approach may be the only way we can ever tackle such questions as, why does space have three dimensions, or why does the proton have a mass that is much larger (precisely 1836 times larger) than the electron, or why is the neutron just 0.14 percent heavier than the proton? If none were the case, we certainly would not be here. One can take the argument further. Perhaps our actual existence requires the universe to have had three space dimensions and the proton mass to be 1836 electron masses. This conclusion is called the anthropic cosmological principle: namely, that the universe must be congenial to the origin and development of intelligent life. Of course, it is not an explanation, and the anthropic principle is devoid of any physical significance.[296]

Not all scientists see much value in the anthropic principle. Owen Gingerich of the Harvard-Smithsonian Center for Astrophysics believes that scientists employ the weak anthropic principle to avoid the conclusion of purpose or design:

> . . . they have turned the original argument on its head. Rather than accepting that we are here because of a deliberate supernatural design, they claim that the universe simply must be this way *because* we are here; Had the universe been otherwise, we would not be here to observe ourselves, and that is that.[297]

John Polkinghorne agrees:

> The Weak Anthropic Principle amounts to little more than tautology. "We're here and so things are the way that makes that possible." It fails adequately to encapsulate the remarkable degree of "fine-tuning" involved in spelling out the conditions that have permitted our evolution. Only a tiny fraction of conceivable universes could have been the homes of conscious beings.[298]

As admitted by Joseph Silk, this anthropic principle offers no explanation how all the perfect conditions were arranged prior to or at Planck time to allow for the subsequent formation of the universe in a precise manner which would allow for the formation of life. The anthropic principle is very uninformative. It can be used to explain anything, but never to give a prediction. Alan Guth writes:

> The anthropic principle is incredibly vague. You can use the anthropic principle, if you want, to explain almost anything. And it never gives precise predictions; it only explains after the fact that what you saw was, in some sense, acceptable. My point of view is that the anthropic explanation is always the resort of last recourse. If you can't find any intelligent theory that's compatible with what you see, that will predict what you see, then you might, as a last resort, entertain a purely anthropic explanation.[299]

Heinz Pagels also has little regard for the principle:

> Physicists and cosmologists who appeal to anthropic reasoning seemed to me to be gratuitously abandoning the successful program of conventional physical science of understanding the quantitative properties of our universe on the basis of universal physical laws. Perhaps their exasperation and frustration in attempting to find a complete, quantitative account for the cosmic parameters that characterize our actual universe has gotten the better of them.[300]

The weak version of the anthropic principle tries to explain the universe's fine tuning by saying that our human existence places us in a privileged time and place. But this "explanation" explains nothing about how all the perfect conditions were set up at the beginning of our universe. Without these carefully selected conditions, no time or place could ever produce or sustain intelligent life. The anthropic principle underestimates how rare—in fact, how impossible—this privileged time and place must be, unless the allowed time or the number of regions is infinite. But time is not infinite. The Big Bang shows that time had a beginning. Our universe was born with natural laws that do not appear to change. Thus these coincidences are not aided by adding more time. And for those that might be helped with more time, the time our universe offers is not nearly enough.

To avoid the problem of the lack of time available from the Big Bang forward, many proponents of accidental development of life propose the Strong Anthropic Principle. This principle contemplates some vague escape around time by hypothesizing multiple universes and infinite multiple universes. Roger Penrose and Stephen Hawking proved that the formation of the universe from the Big Bang was the opposite equivalent to the collapse of a black hole. John Gribbin notes that many physicists have hypothesized from that proof that the collapse of a black hole results in a "bounce" which forms a new universe which expands in a direction away from our universe with a different set of dimensions than our four. Consequently, these physicists believe that our universe is only one among many.[301]

This theory of multiple universes commits the logical fallacy of *petitio principii* if there are only a finite number of universes. However, if the number is infinite, then everything happens, including this universe. The theory of infinite multiple universes allows anyone to argue that somewhere he or she plays basketball better than Michael Jordan and golf better than Tiger Woods. One may propose an infinite array of universes, but a belief in such a proposal rests on faith alone.[302] From some perspectives, this theory violates the principle of Ockham's razor: one should avoid unnecessary assumptions in formulating hypotheses. The razor cuts against the theory of multiple universes because it multiplies the hypothesis. The anthropic principle does not address the question why this universe which is compossible with life exists among the many possible universes which we can conceive which would not be compossible with life. The fine tuning required for life in this universe demands an explanation. To shrug one's shoulders and say, "Well, that is just the way life is," is not consistent with the scientific method which requires that a hypothesis must be susceptible to disproof. The anthropic principles are metaphysical in structure, not scientific. Faith is the essence of the foundation behind these principles, not science. Yet proponents of the anthropic principles are not complacent when alternative metaphysical explanations, such as intelligent design, are offered. Former Cambridge physicist John Polkinghorne notes that the fine tuning of this universe calls for some explanation. He describes John Leslie's firing squad example to illustrate the two principal ways in which humans approach the issue of the universe's fine tuning:

> I am due for execution by fifty crack marksmen. As the sound of firing dies aways, I find that I am still alive. Here is a fact that calls for explanation. It is not enough to say, "Here I am, and that was certainly a close run thing." There are really only two kinds of rational explanation of my good fortune: either there were a very great number of such executions and by lucky chance mine was the one in which they all happened to miss, or the marksmen are on my side. These two lines of thought correspond to two ways people have sought to understand the particularity of our potent universe. . . [303]

PART VI

ETHICAL IMPLICATIONS OF CHANCE OR IMPERSONAL BEGINNING

Although beyond the scope of the questions presented, I will address the ethical implications of a world view with accident or any impersonal cause for the formation of the universe and the formation of the first living matter. Such a world view presents very difficult problems in constructing a foundation for ethical behavior.

The question whether the beginning of time and space was impersonal and the product of accidental or chance processes is an essential question in ethics. If one holds the view of an impersonal beginning, one cannot really talk about what is right or wrong. If the universe was an accident, there are no absolutes, and without absolutes, as Plato stressed, morals do not exist. Right and wrong have the same meaninglessness.

Frederick Nietzsche was one philosopher who held the view of an impersonal beginning, rejected any universal truth, and believed in an absurd world. Nietzsche rejected the concepts of universal right and wrong and asserted that man must decide what is right or wrong by his own will. He called for the emergence of the overman (no relation; my ancestors left Saxony in 1674—200 years before Nietzsche arrived) who creates his own values and defines his own culture. The overman realizes that life is a contest and the enhancement of power is his ultimate purpose. It is easy to see why Nietzsche (in a distorted form) was adopted as the philosopher of National Socialism in Germany.

Nietzsche became insane in 1889 and remained in that condition until he died. I do not know the cause of his insanity, but some of his statements cause me to wonder if his philosophy contributed to his mental condition. Consider Nietzsche's statement from a paragraph he wrote in 1882 entitled, *"What Belongs to Greatness"*:

> Who can attain to anything great if he does not feel in
> himself the force and will to inflict great pain? The
> ability to suffer is a small matter: in that line, weak
> women and even slaves often maintain masterliness.
> But not to perish from internal distress and doubt
> when one inflicts great suffering and hears the cry to
> it—that is great, that belongs to greatness.[304]

Is this statement a valid definition of greatness? If there is no
universal right or wrong, why not? If Nietzsche is correct that
each of us can decide what is right or wrong by our own will,
what is incorrect about his definition of what belongs to greatness?

Despite philosophical assertions to the contrary, no one acts
consistently with a belief that there is no universal right or
wrong. To pay lip service to the absence of right or wrong is one
thing; to be faithful in the practice is quite another. Everyone acts
at times as *conscia mens recti*.[305] Nietzsche said that a personal God
was no longer available for modern man, but without such a God,
all meaning dissolves into absurdity. There is no real basis for
ethics; everything ultimately merges into chaos. Francis Schaeffer
understood this in 1976 when in Switzerland he wrote these
words:

> Without the infinite-personal God, all a person can
> do, as Nietzsche points out, is to make "systems." In
> today's speech we would call them "game plans." A
> person can erect some sort of structure, some type of
> limited frame, in which he lives, shutting himself up
> in that frame and not looking beyond it. This game
> plan can be one of a number of things. It can sound
> high and noble, such as talking in an idealistic way
> about the greatest good for the greatest number. Or it
> can be a scientist concentrating on some small point
> of science so that he does not have to think of any of
> the big questions, such as why things exist at all. It
> can be a skier concentrating for years on knocking
> one-tenth of a second from a downhill run. Or it can
> as easily be a theological word game within the struc-

ture of the *existential methodology*. That is where modern people, building only on themselves, have come,
and that is where they are now.[306]

If Nietzsche is correct, if there is no God, and if the beginning
of the universe was accidental or impersonal; then everything is
ultimately the same: evil is not evil and good is not good. For
Jewish, Islamic, and Christian theists, good is not the same as evil.
A tortuous murder is not the same as a warm embrace. The very
fact that one sees wrong and distinguishes it from right means
one rejects an impersonal beginning to the universe. For Jewish,
Islamic, and Christian theists, God is the moral absolute of the
universe.

PART VII
SUMMARY AND CONCLUSION

7.1. Questions presented

• *The probability of chance causing the formation of a universe compossible with life and the formation of the first form of living matter from inert matter is less than mathematical impossibility at the accepted standard of one in 10^{50}.*

The first question presented in this book was: under standard probability definitions, is it mathematically possible that accidental or chance processes caused (a) the formation of the first form of living matter from non-living matter and (b) the formation of a universe compossible with life? The second question presented was: are current self-organization scenarios for the formation of the first living matter plausible? Most mathematicians normally regard anything with a probability of less than one in 10^{50} as mathematical impossibility. The probabilities calculated under the requirements of molecular biology demonstrate mathematical impossibility for the proposition that accidental or chance processes produced the first living matter. Similarly, the probabilities of the precision of values in particle astrophysics required for the formation of such a universe by accident are too vanishingly small to be considered mathematically possible. The problem in self-organization scenarios is in their failure to distinguish between order and complexity and in the absence of a plausible method of generating sufficient information content into inert matter.

• *Even setting aside the question of mathematical impossibility, an objective, reasonable person following the principles of the scientific method will not favor a proposition with a very low probability over a proposition with a high probability.*

An objective, reasonable person who follows mathematical and other logical thought processes and the principles of the scientific method will not favor a proposition which has a very low probability over a proposition which has an extremely high probability. Because a person's metaphysical assumptions frequently influence his or her interpretation of data, many otherwise rational persons make unwarranted conclusions, which are based not on evidence and logic, but are made in the absence of evidence and contrary to mathematical probabilities, because of their *faith* in the ideology of materialism. *Credo quia absurdum est.*[307] Their conclusions are actually products of their faith in the ideology of materialism, because the selection of a low probability proposition without evidence is not an objective exercise consistent with the methods of science.

7.2. Case against accident from probabilities in molecular biology

• *The central distinction between living and non-living matter is the level of complexity found in the information content of living matter; the Miller and Urey line of experiments does not "work," and the Miller and Urey underlying assumptions are incorrect.*

The central distinction between living and non-living matter is the level of complexity which is found in the information content required to replicate and maintain the organism. In living matter this information content is found in the genetic code and functions of DNA, RNA, and protein synthesis. The theoretical model for the emergence of life by chance processes proposed by Oparin and Haldane and used as the underlying assumptions in the Miller and Urey line of experiments is a failed paradigm. The Miller and Urey line of experiments does not "work." There are factitious flaws in these experiments and in their underlying assumptions, including a less reducing atmosphere for the early earth and an inefficacy in the random distribution of left and right handed molecules (only left handed amino acids are contained in biologically functional proteins). In addition, the dilution processes in any prebiotic soup would have prevented the formation of polypeptides. The existence of the prebiotic soup is crucial to the whole scheme, but there is a complete lack of evidence for the hypothesized soup, a remarkable consideration in the light of college textbooks which present the soup paradigm as an established fact. The Miller and Urey line of experiments is factitious in that it is a series of contrived manipulations by human beings who use the full level of their scientific skills to attempt to form amino acids. These experimenters, using all of their technical skills and intelligence, have failed to produce anything that even remotely resembles a living organism.

• *The period of time available for life to form on earth was only approximately 130 million years which shortens the number of trials allowed in probability calculations and is not sufficient time for chance to be the cause of life.*

The period of time available on earth for the formation of life from accidental or chance processes is exceedingly short. Prior to 3.98 billion years ago the earth was too hot for the emergence of life and was bombarded by meteors. Fossil records, however, show that life was present approximately 3.85 billion years ago or almost immediately after sufficient cooling to approximately 100°C. In other words, only approximately 130 million years were available for chance processes to produce life. This is a much smaller period of time than the billions of years discussed by many proponents of accidental or chance processes.

• *Many different scientists' calculations demonstrate that the formation of life by accidental processes was mathematically impossible.*

Proponents of the formation of life through accidental processes rarely perform the mathematical calculations of the probabilities which lie at the foundation of their hypothesis. Time is the enemy of the occurrence of the unlikely event. Quantitative probabilities demonstrate mathematical impossibility even in a time frame of 15 billion years. The probability of the random formation of a bacterium by chance, as computed by Sir Fred Hoyle and Chandra Wickramasinghe, is one in $10^{40,000}$. Hubert Yockey improved upon the methods of Hoyle and Wickramasinghe's calculations and computed the probability of random, unguided processes generating only a single molecule of the protein iso-l-cytochrome c. The probability calculated was one in 2×10^{-44}, which is also mathematical impossibility. We reviewed similar calculations from other scientists, including Harold Morowitz, who computed the probability of unguided, random development of a single celled bacterium with odds of $10^{100,000,000,000}$ to one. We also discussed the expectation probability for the nucleotide sequence of a bacterium and Bernd-Olaf Küpper's conclusion that even if all the matter in space consisted of DNA molecules of the structural complexity of the bacterial genome, with random sequences, the chances of finding among them a bacterial genome would still be completely neglible.

7.3. Self-organization scenarios and the problem of complexity in information content

• *The problem in self-organization scenarios is in their failure to distinguish between order and complexity and in the absence of a plausible method of generating sufficient information content into inert matter.*

Faced with these odds in an equilibrium system, many scientists have abandoned accident or chance processes and emphasized that the probabilities are greater when considering an open system with an energy source maintaining the system far from equilibrium and from the disorder which inexorably occurs pursuant to the Second Law of Thermodynamics. The problem for these proponents, however, remains one of finding a plausible method of generating sufficient information content into inert matter. The genetic code is the impediment, and theories emphasizing an open system with energy flow do not give a plausible method for directing the energy into the work necessary to form the quality of information content or complexity found in life.

Knowledgeable people fail to distinguish order from complexity. Living systems have complexity in their information content. A system far from equilibrium may spontaneously generate order but not the complexity of information content required in living structures. The theories concerning the generation of order in a system far from equilibrium (such as the theories of Prigogine, Cairns-Smith, Wächtershäuser, Morowitz, or Kauffman) fail because they describe a scenario for the formation of order rather than for complexity and fail to present a plausible scenario for the generation of information content into inert matter.

Even some of the best scientists frequently confuse the concepts of order and complexity. In *The New York Times*, Sunday, September 8, 1996, George Johnson wrote an article describing the patterns and designs caused by jiggling a layer of sand at just the right rhythm. His emphasis was on the spontaneous generation of intricate patterns. He speculated that with the right rhythms scientists might jostle molecules to form cells, and cells join with other cells to produce some weird artificial life. Referring to complexity theorists in the article, Johnson used complexity and higher order as synonyms. This emphasis on order is *nihil ad rem* to the origin of life. Complexity, as defined in this book, is the issue. Systems far from equilibrium can create order, and chaos can give rise to patterns, but that is not complexity. As mentioned at the outset, I am using complexity univocally with the precise definition given by information theory where complexity relates to the level of a structure's information content which in turn is a measure of the minimum number of instructions necessary to specify the structure. To construct a plausible theory for the origin of life, scientists need to discover a theory which explains the generation of complexity, not the generation of order. In terms of the formation of life, information content, such as found in the genetic code, is the stumbling block.

All the proponents of the origin of life by self-organization scenarios named above reject chance origin of life scenarios. We discussed several of the different self-organization theories and found them wanting because they did not present a plausible method of generating sufficient information content in the time available. Again, we noted that the fundamental distinction between living systems is specified complexity, not simple, periodic order. The DNA sequence is highly irregular and aperiodic, similar to letters in a written communication. A crystal, on the other hand, has a simple, periodic repetitive order with very few instructions required to specify its structure. Ilya Prigogine and A. G. Cairns-Smith's theories ignore the vast chasm between simple instructions required for systems such as crystalline order and the extraordinary number of instructions contained in DNA. Crystals do not present a viable explanation for the necessary mechanism, since they represent order with low information content. RNA is also not a sufficient mechanism because the odds against the spontaneous formation of RNA with the capacities of DNA and proteins is no more probable than the formation of the genetic code itself.

• *Life transcends the laws of physics and chemistry and is not reducible to these laws; novel approaches to the origin of life have not answered the fundamental questions raised by complexity.*

Harold Morowitz rejects Jacque Monod's chance origin of life scenario. He, like many proponents of self-organization scenarios, considers deterministic processes consistent with a religious point of view and regards the formation of closed vesicles to be a major event in the origin of life. He rejects the hypothesis of clay or pyrite-related scenarios because they violate the principle of continuity. His theory differs from others in that he proposes that the genetic code is a later event in the origin of life process. He has structured a novel approach, but has not solved the questions raised by information theory or explained a method for generating sufficient information content into inert matter. The information generation is not likely to flow from the laws of physics or chemistry alone, because the genetic information content of the genome, for constructing even the simplest organisms, is much larger than the information content of these laws. In addition, a law produces regular, predictable patterns. But a repeating pattern encodes little information. The information in a DNA molecule is flexible and independent of the base of sugars and phosphates which comprise the molecule. Because the information is independent from these chemicals, the information did not arise from the chemicals; just as the words in this book did not arise from the ink in my computer printer.

Stuart Kauffman also rejects the chance origin of life scenario. His proposals of complexity arising on the edge of chaos are reminiscent of Prigogine's confusion of complexity with order. He relies to a great extent on computer simulations, reducing organisms to mathematical symbols which are then manipulated. His reliance on computer simulation is criticized by Morowitz and by Stanley Miller, who are advocates of laboratory experiments. The vexing question for him, as for others, is the formation of the genetic code.

7.4. ALH84001

• *The meteorite ALH84001 is not significant evidence for spontaneous formation by chance processes.*

The recent discovery of possible remnants of bacterial life on a meteorite from Mars is very controversial. The evidence for life centers on four findings which each have organic or inorganic explanations. The carbonate globules, magnetite, iron sulfide and PAHs are all frequently generated in large quantities by inorganic and organic processes. PAHs are present throughout the Milky Way from inorganic processes, and magnetite can be formed by inorganic precipitation and other processes. The pyrite in ALH84001 contains extra heavy sulfur and is inconsistent with biological activity. Mars and earth have exchanged an enormous tonnage of meteorites for billions of years. The presence of micro-organisms on Mars is not significant evidence for spontaneous generation by chance processes. Meteorites and the solar winds have moved vast amounts of matter from the earth to Mars and throughout the solar system over the past four billion years. For purposes of the question presented, it is not crucial whether life formed on Mars or whether life was first transported to Mars from earth; the probabilities are against chance processes under either condition.

7.5. The necessary bridge

• *The necessary bridge which science must cross to pass over the abyss separating the physico-chemical world from biology is the origin of complexity found in the information content of the genetic code; at present, no valid scientific explanation of the origin of life exists.*

On the basis of the mathematical probabilities presented in this book one would need to make a long leap of faith to accept the proposition of the origin of life from accidental or chance events. This is certainly a longer leap than the leap of faith required to consider the possibility of life emerging from the work of an intelligent designer. The question frequently asked to refute the argument for a designer is: if there was a designer, who made the designer? This question is not valid in an application to a supreme being as designer, because it assumes a self contradiction, i.e., that the designer was designed. This is similar to asking the question: who or what made triangles circular? The argument that a designer needs a designer is only valid from the anthropomorphic perspective of a designer subject to time. Anyone or anything, including a designer, subject to time, would have a beginning. If something has a beginning, it has a cause. But the designer need not be confined to time. If the cause of the universe must come from outside the dimensions in the universe, the possibility exists that the designer would not have a beginning and would not "need a designer." Such a designer would be outside of time and would have no beginning and no end.

The mathematics are overwhelmingly against accident, and the evidence for self-ordering in the elements of the periodic table or otherwise has not been demonstrated. No valid scientific explanation for the origin of life exists at the present time. The most that can be said is that because the mathematics require a very long leap of faith over a chasm to arrive on the mythical side of life emerging by accident, a rational person would not make that leap; the odds against success are too vast. Origin of life scenarios based on accidental processes proceeding pursuant to the theories hypothesized by the Miller and Urey line of experiments are not only highly improbable; they are mathematically impossible given the finite time available for such processes to form the first living matter.

7.6. Case against accident from probabilities related to precision of values in particle astrophysics

• *The probabilities against the accidental formation of a universe compossible with life compounds the probabilities against the accidental formation of living matter. Because the universe is finite in spacetime and had a beginning, everything is not necessarily possible; probability calculations must be made in the context of finite spacetime boundaries.*

The evidence and mathematical probabilities against accident in the formation of the first form of life from inert matter is compelling. At least equally compelling are the evidence and mathematical probabilities against the accidental formation of a universe compossible with life. Because the formation of life necessarily requires both the formation of a universe compossible with life and the formation of the first form of life from inert matter, when one considers the probability of the existence of life, the odds against accident are compounded by considering the probabilities against the formation of such a universe coupled with the probabilities against the formation of the first form of life from inert matter.

The question whether the universe had a beginning is important in calculating mathematical probabilities. In an infinite, ageless universe, anything can happen. Accordingly, we began the section of the book concerning the precision of values in particle astrophysics with a background discussion starting with Hubble's discovery of an expanding universe and the confirmation of that expansion by the COBE satellite measurements of the cosmic background radiation. An expanding universe implies that the universe was previously smaller. Reversing the rate of expansion compresses all of the matter in the universe in an infinitely dense singular point smaller than a proton. To understand this singularity, we compared the effects on space and time caused by a black hole. To understand this compression and provide a background for our discussion of the precision of values in particle astrophysics, we reviewed briefly the four fundamental forces and the structure of quantum particles, including the subatomic interactions at the level of quarks, leptons, gluons, and other members of the particle garden. We discussed the grand unified extra-dimensional theories which could assist in answering many physical and metaphysical paradoxes. The forces and particles in the atomic and subatomic world displayed an early universe with remarkable symmetry, elegance and precision necessary for the formation of an environment compossible with living matter.

• *The remarkable precision in values of particle astrophysics, including the fundamental constants, the strength of the four forces, and the mass of the elementary particles, disclose a fine tuning required for a universe compossible with life.*

With this background in particle astrophysics and the activity of the forces and particles in the early universe, we began a review of many examples of the fine tuning of these forces and values which were a prerequisite for the formation of life in this universe. We first reviewed the resonance precision required for the formation of carbon, a key element of life. Sir Fred Hoyle admitted that his atheism was dramatically disturbed when he calculated the odds against the precise matching required to form a carbon atom through the triple alpha process. He said the number one calculates from the facts is so overwhelming as to put the conclusion that a superintellect had monkeyed with physics almost beyond question. No less impressive was Paul Davies's calculation that the matching of the explosive force of the Big Bang and gravity was one part in 10^{60}, a precision that was equal to the odds of a random shot of a bullet hitting a one inch target from a distance of twenty billion light years.

Closely related to the fine tuning of the expansion rate is the precise matching of the density of matter after Planck time with the critical density. The slightest deviation in this matching would have made life in the universe impossible. This density had to match critical density to more than 50 decimal places. Although a model for an inflationary epoch in the Big Bang could lead to such a matching, the inflationary epoch itself would have to be fine tuned to provide for this astounding precision.

The strong force which binds the particles in an atom's nucleus must be balanced with the weak nuclear force to a degree of one part in 10^{60}. If the strong force were any weaker, atomic nuclei could not hold together and only hydrogen would exist. If the strong force were only slightly stronger, hydrogen would be an unusual element, the sun would not exist, water would not exist, and the heavier elements necessary for life would not be available.

The electromagnetic force must be precisely balanced with the gravitational force to allow for the formation of a star such as the sun. If gravity's force were changed by only one part in 10^{40}, stars like our sun would not exist but only stars either too hot or cold to support life on any portion of a surrounding planetary system. In the balance between the electromagnetic force and the force of gravity, the number of electrons must be meticulously balanced to an accuracy of one part in 10^{37} with the number of protons. Without such a precise balance stars and planets would not have formed.

Any deviation in the strength of the electromagnetic force would also preclude molecular formation necessary for life. The electromagnetic force must be precisely balanced with the ratio of electron mass to proton mass. The proton is 1,836 times heavier than the electron. This is a fundamental ratio which is very finely adjusted to make possible the development of life. Moreover, the mass of the proton and the mass of the neutron are meticulously balanced. The emergence of life depended on an astounding precision among the masses of these three particles. Even the slightest variation would prevent the formation of living matter.

The Second Law of Thermodynamics requires that entropy or disorder in the universe tends toward a maximum. Because the universe could not have been dissipating from infinity or it would have run down, it must have had a beginning and a very highly ordered beginning. If the Big Bang is regarded as only an impressive accident, there is no explanation why it produced a universe with such a high degree of order, contrary to the Second Law. In a 1979 computation Roger Penrose calculated that the probability of the observed universe occurring by chance was one in 10^{300}. In another calculation in 1989 Penrose computed that to provide a universe compatible with the Second Law the precision needed to set the universe on its highly ordered course was to an accuracy of one part in $10^{10^{(123)}}$. This number is so large that even if a "0" was written on every proton and neutron (and every other particle) in the known universe, the universe would not contain enough matter to write down the figure.

The weak nuclear force is also highly fine tuned. If the force were slightly larger, no helium would form and no heavier elements necessary for life would exist. If the force were slightly weaker, no hydrogen would be available and life would not be possible. Supernova explosions which spread the heavier elements necessary for life depend on a very precise value in the weak nuclear force.

The laws of physics and chemistry are only compatible with the emergence of life in no more than three spatial dimensions. These laws, such as Pauli's exclusion principle, allow the DNA molecule to convey the information necessary for life and maintain the fixed properties that underlie the genetic code.

An accidental universe cannot explain the astounding agreement between abstract mathematics and the laws of the physical world. Abstract mathematics have predicted counter intuitive phenomena to a remarkable precision. The agreement between the counter intuitive theory of general relativity and the physical world has been confirmed by experiment to more than a trillionth percent precision. Precision to this degree cannot be explained by chance alone. Similarly, the strange unseen, counter intuitive subatomic world of quantum theory matches the predictions of abstract mathematics to a remarkable degree. Our minds seem to be finely tuned to the structure of the universe. This fine tuning cannot be understood as a curious spin-off from the need of our ancestors to dodge a wild animal.

The abundance of evidence from the precision of values in particle astrophysics against the accidental formation of a universe compossible with life is difficult to overstate. Force strengths, particle masses and other physical values are accurate and precise, not for just one reason, but for two, or three or five. The fine tuning is not just between particles, forces and values, but among many factors so that the permutations of precision and fine tuning is compounded and immense.

- ***The theory of an oscillating universe provides no escape from these probability calculations.***

Because the mathematical probabilities against accident are so overwhelming in our universe, some scientists are attracted to the concept of an oscillating universe which allows for an infinite number of beginnings. Infinity can be used to explain almost anything so anyone displeased with the mathematical calculations against accident may grasp at any opportunity to bring infinity into the examination. Stephen Hawking and Roger Penrose, however, have demonstrated that the gravitational force on a collapsing universe would produce a singularity which would not allow for another expansion. A Big Crunch would be totally chaotic and the entropy calculated at the Crunch would be so large that it would preclude another expansion.

• *Some theorists speculate that a quantum fluctuation caused the formation of the universe under unknown conditions; however, the dynamics of quantum mechanics in conventional space and time would not allow for more than a fleeting duration for the universe; moreover, even assuming that the universe was formed from a quantum fluctuation, the question of the origin of the laws of physics and the quantum field remains and indicates that a logic or intelligence existed prior to spacetime.*

Because the development of something out of nothing requires a cause, some physicists have asserted that the entire universe came into being by a quantum energy fluctuation. This proposal is made in accordance with Heisenberg's uncertainty principle. The problem with the proposal is that any "virtual" particle could borrow energy for its appearance for only a fleeting duration. The shorter the duration, the larger the potential virtual particle. Since the amount of time for the duration of these particles varies inversely with their mass, the duration for a system with the mass of the universe would be well short of Planck time, not to mention the billions of years required by the estimate of the age of the universe computed by any reasonable figure for Hubble's constant. Moreover, because the size of the time interval is related to the probability of a quantum event, at time zero the probability of a quantum fluctuation would be zero.

Such a speculation does not answer the question of the origin of the laws of physics, including quantum mechanics and the origin of the field from which the fluctuation could have occurred. Even assuming the validity of the speculation on the formation of the universe from something analogous to a quantum energy fluctuation, we are left not with the answer of accident or chance but with the question of the origin of the laws of physics and the origin of the quantum field. With Heinz Pagels we ask the question, where did these laws come from? Where are these laws written into the void that "tells" the void that it is "pregnant with a possible universe?" Even the void is not subject to accident or chance, but to "a logic that exists prior to space and time."

• *Hawking uses imaginary numbers in imaginary time to remove a singularity (beginning), but he emphasizes that his no*

*boundary concept is not even a theory but only a proposal which
cannot be deduced from any known, verified principle; the proposal
raises certain issues concerning the Second Law of Thermodynamics and the asymmetry of time.*

Some physicists attempt to avoid the requirement of the universe's beginning. In 1970 Roger Penrose and Stephen Hawking
published a paper which proved that in an expanding universe
with as much matter as we observe and where the theory of general relativity applied, an initial singularity must have existed.
The beginning of time was a point of infinite density and infinite
curvature. Subsequently, Hawking and James Hartle published a
paper where a singularity might be avoided by a quantum mechanical wave function provided certain assumptions were made
which are contrary to our understanding of the quantum state of
the universe. Hawking later speculated that a singularity could
be avoided if imaginary numbers were used in imaginary time.
We discussed the Hawking and Hartle paper and Hawking's subsequent proposal (which he emphasizes is only a proposal and
not a theory) using imaginary time and noted that when Hawking returns to the real time in which we live, the universe has a
beginning and an end. This conclusion fits well with the application of the Second Law which also requires a beginning and an
end to the universe.

•   *The Weak Anthropic Principle is little more than a tautology
with no predictive value, and the Strong Anthropic Principle begs
the question and ignores the principle of Ockham's razor.*

Something happened prior to Planck time but the equations of
science cannot speak to it; time and space cannot even be part of
it. The question remains: how can something arise from nothing?
The weak anthropic principle amounts to little more than a tautology, fails to account for the remarkable degree of precise values in the universe, is of little predictive value, and does not
explain anything. The principle is incredibly vague and can be
used to assert almost any hypothesis. In certain respects, the
strong anthropic principle violates the principle of Ockham's razor. If the number of universes is infinite, then everything hap-

pens, including this universe, and in some other universe I can play basketball like Michael Jordan and golf like Tiger Woods. One may find multiple universes easier to believe in than one finely tuned universe, but that belief rests on faith alone.

• *Any world view consistent with accident or any other imper-sonal cause for the formation of the universe and the first living matter inexorably leads to relativistic ethics which permit anyone to define right or wrong by personal will.*

Although beyond the scope of the questions presented, we brief-ly noted that any world view consistent with accident or any oth-er impersonal cause for the formation of the universe and the first form of living matter inexorably leads to a rejection of the concepts of universal right or wrong. Anyone can then decide what is right or wrong by one's own personal will. This is the po-sition of Frederick Nietzsche, who defined greatness as the will to inflict great pain on one's fellow human beings without any in-ternal distress within one's self. To pay lip service to the absence of universal right and wrong is one thing, but no one lives consis-tently with a belief that right and wrong is dependent on a per-son's will. Everyone acts at times as *conscia mens recti.*

• *The central conclusions from this analysis are: accident is not a mathematically possible explanation for the finely tuned particle physics required for a universe compossible with life or for the mo-lecular origin of life; all self-organization scenarios fail to explain the mystery of information generation; and life transcends the laws of physics and chemistry.*

In conclusion, accident is not a mathematically possible explana-tion for the formation of a universe compossible with life. The mathematical odds against an accidental universe are well be-yond the standard definition of a probability of less than one in 10^{50}. The emergence of life required an astounding fine tuning of the particle physics of the early universe. As noted above, Roger Penrose computed that the probability of the observed universe occurring by chance is one in 10^{300}. Other calculations demon-strate an even more remote probability. In addition, the fine tuning is exactly what is required not just for one reason, but for two or three or five reasons. It is not plausible that a particle mass or a force strength could be tuned in first one way and then another to satisfy several conflicting requirements for the development of life. Moreover, accident is not an adequate explanation for the exceptional performance of counter intuitive abstract mathematics reflecting the real structure of the physical universe. Where do the laws of physics come from? The explanations of the weak and strong anthropic principles are insufficient in that they are metaphysical in nature, constitute little more than a tautology with no predictive value, beg the question if the number of multiple universes is not infinite, and are incredibly vague.

Because the formation of life requires the formation of a universe compossible with life, the case against accident as an explanation for life is satisfied completely by an examination of the probabilities involved in the fine tuning of particle astrophysics without regard to the issues raised by molecular biology. When one *couples* the probabilities in physics against an accidental universe compossible with life with the molecular biological and pre-biological probabilities against the formation of the first form of life from inert matter, the compounded calculation wipes the idea of accident entirely out of court.

Some theorists who recognize the mathematical impossibility associated with accident as the cause of life speculate with self-organization scenarios. At present, however, all self-organization scenarios fail in explaining the generation of sufficient information content to qualify a structure as a life form. Because life transcends the laws of physics and chemistry and is not reducible to these laws, an adequate self-organization scenario may never be discovered. The information generation in living matter is not likely to flow from the laws of physics or chemistry alone, because the genetic information content of the genome, for constructing even the simplest organisms, is much larger than the information content of these laws. In addition, a law produces a regular, predictable pattern, but repeating patterns encode little information. The information in a DNA molecule is flexible and independent of the base of sugars and phosphates which comprise the molecule. Because the genetic information is independent from these chemicals, the information did not arise from the chemicals; just as the words in this book did not arise from the ink in my computer printer.

Life cannot be explained by an appeal to accident; and if life transcends the laws of physics and chemistry, then the origin of life will never be demonstrated by an adequate self-organization scenario, but remain an intractable or indeterminate problem such as that represented by Gödel's Incompleteness Theorem in mathematics. If life transcends the laws of physics and chemistry, the cause of life is more than a physical thing. At this point we enter into metaphysics ("meta" means beyond), but it is worth noting that an entity which is more than a thing is a person. If life transcends the laws of physics and chemistry, then a rational conclusion is that a Person, not chance and the laws of physics and chemistry, caused and is causing life.

APPENDIX
SOME IMPORTANT PHYSICAL VALUES

A. PHYSICAL CONSTANTS

Speed of light $\qquad$ $c = 2.99792458 \times 10^8$ m/s

Gravitation constant $\qquad$ $G = 6.668 \times 10^{-11}$ Nm^{2}kg^{-2}

Planck constant $\qquad$ $h = 6.626076 \times 10^{-34}$ J·s

$\qquad$ $= 4.1355670 \times 10^{-15}$ eV·s

$\qquad$ $h_p = h/2\pi = 1.054573 \times 10^{-34}$ J·s

$\qquad$ $= 6.582122 \times 10^{-16}$ eV·s

Boltzmann constant $\qquad$ $k = 1.38066 \times 10^{-23}$ JK^{-1}

Charge of electron $\qquad$ $e = 1.6021773 \times 10^{-19}$ C

Mass of H atom $\qquad$ $m_H = 1.673 \times 10^{-27}$ kg

B. SOME PARTICLE MASSES

	kg
Electron	$9.1093897 \times 10^{-31}$
Proton	$1.6726231 \times 10^{-27}$
Neutron	1.674955×10^{-27}
Deuteron	3.343586×10^{-27}

C. CONVERSION FACTORS

1eV $= 1.6021773 \times 10^{-19}$ J

1 light-year $= 9.46 \times 10^{15}$ m

1 parsec $= 3.26$ light-year

0 degrees Kelvin = -273 degrees Centigrade =
-460 degrees Fahrenheit

D. SOLAR MASS

$M\odot = 2 \times 10^{30}$ kg

NOTES

1. Typical size estimates for the atom and some subatomic components are: atom = 10^{-8} cm, nucleus = 10^{-12} cm, nuclear particle = 10^{-13} cm, electron and quark = 10^{-14} cm or less. Quarks are so small that their precise size is not known. Curt Suplee gives the following example to demonstrate how small a quark actually is: ". . . if an atom were the diameter of the District of Columbia, its nucleus would be no bigger than a bushel basket of hollow softballs, each of which represents one proton or neutron. Suspended inside every ball would be three invisibly minuscule specks which—if they have any measurable size at all, which they may not—would occupy no more than one-billionth the volume of the whole ball. Those are the quarks." Curt Suplee, "Gently Nudging the Mystery Out of the Reclusive Quark", *The Washington Post*, p. A3, April 28, 1997.

2. The concept of life originating spontaneously from inorganic matter.

3. As my White House Fellow colleague, Don Haider, now a Professor at Northwestern University, simply and frequently noted, "Where one stands, depends upon where one sits."

4. We readily believe what we wish to believe.

5. George Gamow, as quoted in Ronald Clark, *Einstein: The Life and Times* (New York: World Publishing Company, 1971), p. 215.

6. Inference from the universal to the particular is valid.

7. Inference from the particular to the universal is not valid.

8. Lewis Carroll, *Alice in Wonderland* (New York: The New American Library, Inc., 1960), pp. 67–68.

9. Mark Twain, *Life on the Mississippi* (New York: Bantam Books, 1981), p. 93.

10. George Johnson, *Fire in the Mind* (New York: Alfred A. Knopf, 1995), p. 281.

11. S. Morris Engel, *With Good Reason* (New York: St. Martin's Press, 1976), p. 59.

12. Lewis Carroll, *Through the Looking Glass* (New York: The New American Library, Inc., 1960), p. 196.

13. C. Stanley Ogilvy, *Excursions in Mathematics* (Mineola, New York: Dover Publications, Inc., 1994), pp. 38, 39, 139 and 140.

14. Arthur Benjamin and Michael Brant Shermer, *Mathemagics* (Los Angeles: Lowell House 1993), pp. 161–163.

15. See Harold Parks, Gary Musser, Robert Burton, and William Siebler, *Mathematics in Life, Society & the World* (Upper Saddle River, N.J.: Prentice-Hall, Inc., 1997).

16. Peter Bernstein, *Against the Gods: The Remarkable Story of Risk* (New York: John Wiley & Sons, Inc. 1996), p. XXXVI.

17. Dominic Olivastro, *Ancient Puzzles* (New York: Bantam Books, 1993), pp. 212–215.

18. To illustrate the Fibonacci contrivance when the addition results in a number less than 211 and more than 105, assume that your friend secretly selects the number 56. Dividing 56 by 3, your friend tells you that the remainder is 2. You multiply 2 by 70 with the result of 140. Dividing 56 by 5, your friend tells you that the remainder is 1. You multiply 1 by 21 with the result of 21. Dividing 56 by 7, your friend tells you that the remainder is zero. The sum of the addition of the results of your multiplication is 161 (140 plus 21). Subtracting 105 from 161, you arrive at the contrived number 56.

19. Gordon Fraser, "The Human Paradox: Stephen Hawking and His Work," *Science Spectra* **4**, 13 (1996).

20. G. Polya, *How to Solve It: A New Aspect of Mathematical Method*, 2nd ed. (Princeton: Princeton University Press, 1988), p. 142.

21. Robert Shapiro, *Origins: A Skeptic's Guide to the Creation of Life on Earth* (New York: Simon and Schuster, Summit Books, 1986), p. 112.

22.   Joseph Heller, *Catch-22* (New York: Bantam Doubleday Dell Publishing Group, Inc., 1955, 1994), pp. 39–40.

23.   After this, therefore because of this.

24.   *William Shakespeare, King Henry IV, The Complete Works of Shakespeare*, Harding Craig (ed.) (Chicago: Scott Foresman and Co., 1961), pp. 690–691.

25.   Voltaire, *Candide*, as quoted in Engel, p. 78.

26.   Engel, pp. 78–79.

27.   By equally valid reasoning.

28.   Martin Gardner, *HexaFlexegons and Other Mathematical Diversions* (Chicago: University of Chicago Press, 1988), pp. 52–54.

29.   For a description of these and other paradoxes, including Jean Buridan, the fourteenth century Venetian philosopher's formulation, see Nicholas Faletta, *The Paradoxicon* (New York: John Wiley & Sons, Inc., 1990), pp. 71–78.

30.   Roger Penrose, *The Emperor's New Mind* (New York: Oxford University Press, 1989), p. 58.

31.   Faletta, pp. 70–86; and Rudy Rucker, *Infinity and the Mind* (New York: Bantam Books, 1983), pp. 181-191.

32.   For a readable explanation of Gödel's ideas, see Ernst Nagel and James Newman, *Gödel's Proof* (New York: New York University Press, 1986).

33.   As noted in the table in the Appendix, the Planck constant is equal to 6.626176×10^{-34}Js. J is the symbol for Joule. One Joule = 10^{7}ergs or 0.2388 calorie.

34.   Dear God does not play dice with the world.

35.   John Gribbin, *In the Beginning: After COBE and Before the Big Bang* (Boston: Little, Brown & Company, 1993), p. 46.

36.   George Gaylord Simpson, "The Nonprevalence of Humanoids," *Science* **143**, 771 (1964). This definition is given to emphasize the requirement of information content to consider matter to be alive. Hubert Yockey, the leading authority on information theory and biology, has mathematical objections to the concept of

negenthropy and recognizes the current confusion between order and complexity which in some respects have opposite meanings, the latter not violating the Second Law of Thermodynamics
with random processes in a closed system. See Hubert Yockey,
Information Theory and Molecular Biology (Cambridge: Cambridge
University Press, 1992).

37.   Gribbin, p. 47.

38.   Michael Denton, *Evolution: A Theory In Crisis* (Bethesda,
Md.: Adler and Adler, 1985), pp. 250 and 271.

39.   Jean P. Milani, et. al., *Biological Science* (Lexington, Mass.:
D.C. Heath and Company, 1990), p. 225.

40.   Hubert Yockey, "Information in Bits and Bytes," *BioEssays*
17, 85 (1995).

41.   A. I. Oparin (1924). Proiskhozdenic Zhizny, Moscow: Izd
Moskovski Rabochi. Published in English translation in J.D. Bernad, *The Origin of Life* (London: Weidenfield and Nicolson, 1967).

42.   J. B. S. Haldane, *The Origin of Life*. In *Rationalist Annvel* (London: C.A. Watts & Co., 1929).

43.   Jacques Monod, *Chance and Necessity* (London: Collins,
1972), pp. 110, 137, and 167.

44.   A monomer is a molecule of low molecular weight capable
of reacting with identical or different molecules of low molecular
weight to form a polymer, a compound of higher molecular
weight.

45.   The presence of oxygen in the very early earth's atmosphere can be explained by the process of photodissociation of
water. Ultraviolet light from the sun could produce oxygen in the
primitive earth's atmosphere as follows: $2\ H_2 0 + (hv)$ ultraviolet
light energy $= 2\ H_2 + O_2$.

46.   Denton, p. 261.

47.   R. T. Brinckmann, 1969, *Journal of Geophysical Resources*,
745355 as cited in Charles B. Thaxton, Walter L. Bradley, and Roger Olsen, *The Mystery of Life's Origin: Reassessing Current Theories*
(New York: Philosophical Library, 1984), p. 79.

48. J. H. Carver, "Prebiotic Atmospheric Oxygen Levels", *Nature* **292**, 136 (1981).

49. Thaxton, Bradley, and Olsen, p. 80.

50. Denton, p. 262.

51. Thaxton, Bradley, and Olsen, p. 43.

52. Stanley Miller as quoted in Shapiro, p. 112.

53. Shapiro, p. 112.

54. Shapiro, p. 112.

55. Shapiro, p. 114.

56. A peptide bond is a covalent chemical bond formed between two amino acids. As noted elsewhere, a polypeptide is a long chain of chemically bonded amino acids.

57. Hydrolysis is the process in which a compound is split into other compounds by taking up the elements of water.

58. For a thorough analysis of these dilution processes, see Thaxton, Bradley, and Olsen, pp. 42–67.

59. Michael J. Behe, *Darwin's Black Box* (New York: The Free Press, 1996), pp. 169–170.

60. Hubert Yockey, "Information in Bits and Bytes," p. 87.

61. J. Brooks and G. Shaw, *Origin and Development of Living Systems* (New York: Academic Press, 1973), p. 23.

62. Denton, pp. 260–261.

63. Hubert Yockey, "Comments on Let There Be Life ; Thermodynamic Reflections on Biogenesis and Evolution by Avshalom C. Elitzur," *Journal of Theoretical Biology* **176**, 351 (1995).

64. Yockey, *Information Theory and Molecular Biology*, pp. 235, 236, 238, and 335.

65. Phillip E. Johnson, *Darwin on Trial* (Washington, D.C.: Regnery Gateway, 1991), p. 103.

66. Brooks and Shaw, p. 212. Michael Behe agrees: "There were no chemists four billion years ago. Neither were there any chemi-

cal supply houses, distillation flasks, nor any of the many other devices that the modern chemist uses daily in his or her laboratory, and which are necessary to get good results. A convincing origin-of-life scenario requires that intelligent direction of the chemical reactions be minimized as far as possible. Nonetheless, the involvement of some intelligence is unavoidable. Reasonable guesses about what substances were available on the early earth—such as Stanley Miller made—are a necessary starting point. The trick for the researcher is to choose a probable starting point, then keep his hands off." Behe, p. 168-169.

67.   Hubert P. Yockey, "Self Organization Origin of Life Scenarios and Information Theory," *Journal of Theoretical Biology* **91**, 14 (1981).

68.   George Wald, "The Origin of Life", *Scientific American* **191**, 48 (1954).

69.   Harold J. Morowitz, *Beginnings of Cellular Life* (New Haven: Yale University Press, 1992), p. 31.

70.   John M. Hayes, "The Earliest Memories of Life on Earth," *Nature* **384**, 21 (1996).

71.   Morowitz, p. 30.

72.   Hayes, p. 22.

73.   S.J. Mojzsis, G. Arrhenius, K.D. McKeegan, T.M. Harrison, A.P. Nutmjam and C.R.L. Friend, "Evidence for Life on Earth Before 3,800 Million Years Ago," *Nature* **384**, 55 (1996).

74.   A. G. Cairns-Smith, *Genetic Takeover and the Mineral Origins of Life* (Cambridge, U.K.: Cambridge University Press, 1982).

75.   Morowitz, pp. 36–37.

76.   Ian Stewart, "The Interrogator's Fallacy," *Scientific American* **275**, 172 (1996).

77.   Pierre LeComte du Noüy, *Human Destiny* (New York: Longmans, Green and Co., 1947), pp. 28–29.

78.   William Shakespeare, *Macbeth*, Act V, Scene V.

79. See example of similar calculation in Gerald L. Schroeder, *Genesis and the Big Bang* (New York: Bantam Books, 1990), p. 186.

80. Cesare Emiliani, *The Scientific Companion* (New York: John Wiley & Sons, Inc., 1995), p. 151.

81. Richard Dawkins, *The Blind Watchmaker* (New York: W.W. Norton & Company, Inc., 1986).

82. The question fails; the argument collapses.

83. Yockey, *Information Theory and Molecular Biology*, p. 279.

84. Hoyle and Wickramasinghe, *Evolution from Space* (London: J. M. Dent & Sons, 1981).

85. Hoyle and Wickramasinghe, pp. 148, 24, 150, 30, and 31 as quoted in Thaxton, Bradley, and Olsen, p. 196.

86. "Threats on Life of Controversial Astronomer", *New Scientist*, January 21, 1982, p. 140.

87. See, for example, F. Crick, *Life Itself* (New York: Simon and Schuster, 1981).

88. Morowitz, pp. 4 and 12. William of Ockham (Occam) was a fourteenth century logician, physicist and theologian who wrote in his *Quodlibeta Septem* that "when a proposition comes out true for things, if two things suffice for its truth, it is superfluous to assume a third." In other words, the principle of Ockham's razor calls for the least number of assumptions in the construction of an explanation.

89. Yockey, *Information Theory and Molecular Biology*, p. 257.

90. Walter L. Bradley and Charles B. Thaxton, "Information and the Origin of Life" in *The Creation Hypothesis*, ed. J. P. Moreland (Downers Grove, Il.: InterVarsity Press, 1994), p. 190.

91. Harold J. Morowitz, *Energy Flow in Biology* (Woodbridge, Conn.: Ox Bow Press, 1979), p. 12.

92. Morowitz, *Energy Flow in Biology*, p. 68.

93. Bernd-Olaf Küppers, *Information and the Origin of Life* (Cambridge, Mass.: The MIT Press, 1990), pp. 59–60.

94.   Denton, pp. 328, 329, and 342.

95.   You cannot make Mercury out of just any log.

96.   Denton, p. 331.

97.   Thaxton, Bradley, and Olsen, p. 124.

98.   Thaxton, Bradley, and Olsen, pp. 124–125.

99.   Thaxton, Bradley, and Olsen, p. 125.

100. Nothing to do with the matter. See Thaxton, Bradley, and Olsen, p. 124f.

101. Behe, p. 171.

102. Gerald Joyce and Leslie Orgel, as quoted in Behe, p. 172.

103. Yockey, *Information Theory and Molecular Biology*, p. 282.

104. Robert Shapiro, "Prebiotic Ribose Synthesis: A Critical Analysis, *Origins of Life and Evolution of the Biosphere* 18 (1988).

105. Harold J. Morowitz, *Beginnings of Cellular Life*, p. 4.

106. Yockey, *Information Theory and Molecular Biology*, p. 172.

107. Yockey, *Information Theory and Molecular Biology*, p. 280.

108. Another way must be tried.

109. Yockey, *Information Theory and Molecular Biology*, pp. 245 and 281.

110. Thaxton, Bradley, and Olsen, p. 152.

111. See A. G. Cairns-Smith.

112. Nancy R. Pearcey and Charles B. Thaxton, *The Soul of Science* (Wheaton, Il.: Crossway Books, 1994), p. 238.

113. Yockey, "Self Organization Origin of Life Scenarios and Information Theory", p. 20.

114. You are weaving a rope of sand which, of course, is attempting the impossible.

115. Thaxton, Bradley, and Olsen, p. 131.

116. Yockey, *Information Theory and Molecular Biology*, p. 236. As

discussed in the text, one of the problems of the origin of life scenario in the prebiotic soup is the thermodynamics and kinetics of polycondensation in aqueous solutions where water molecules break the bonds of biochemical polymers in the hydrolysis process and prevent the formation of longer polymers that are necessary to form the genetic system. Recently James Ferris and Leslie Orgel and their colleagues provided evidence that longer polymers can be obtained if the polycondensation occurs on a metallic or clay surface free of aqueous solution. Their results were published in *Nature* **381**, 59 (1996). Although interesting, the experiment does not deal with replication and fails to answer the questions raised by information theory.

117. John Horgan, "In the Beginning . . . ," *Scientific American* **264**, 121 (1991).

118. Horgan, "In the Beginning," p. 122.

119. Horgan, "In the Beginning," p. 122.

120. Horgan, "In the Beginning," p. 122.

121. In the experiment an aqueous slurry of coprecipitated nickel sulfide and ferrous sulfide converted carbon monoxide and methansthiol (CH_3SH) into the activated thioester (CH_3-CO-SCH_3) which hydrolyzed to acetic acid. In the presence of aniline, acetanilide was formed. When the nickel and ferrous sulfides were modified with a catalytic amount of selenium, acetic acid and methansthiol were formed from carbon monoxide and hydrogen sulfide alone. Huber and Wächtershäuser believe that the reaction cam be considered as the primordial initiation reaction for a chemoautotrophic origin of life. Claudia Huber and Günter Wächtershäuser "Activated Acetic Acid by Carbon Fixation on (Fe, Ni)s Under Primordial Conditions," *Science* **276**, 245 (1997).

122. Yockey, *Information Theory and Molecular Biology*, p. 266.

123. An amphiphilic molecule has a polar end and a non-polar end.

124. Morowitz, *Beginnings of Cellular Life*, p. 27.

125. Morowitz, *Beginnings of Cellular Life*, p. 168.

126. Morowitz, *Beginnings of Cellular Life*, pp. 174–178.

127. Johnson, pp. 225–226.

128. Yockey, *Information Theory and Molecular Biology*, p. 335.

129. Pearcey and Thaxton, p. 231.

130. In a later essay entitled, "Life's Irreducible Structure," Polanyi considered a hypothetical DNA molecule which was determined by the laws of chemistry and physics. After describing the limited operations of such a molecule, he concluded that it would not have appreciable information content: "Suppose that the actual structure of a DNA molecule were due to the fact that the bindings of its bases were much stronger than the bindings would be for any other distribution of bases, then such a DNA molecule would have no information content. Its code-like character would be effaced by an overwhelming redundancy. We may note that this is actually the case for an ordinary chemical molecule. Since its orderly structure is due to a maximum of stability, corresponding to a minimum of potential energy, its orderliness lacks the capacity to function as a code. The pattern of atoms forming a crystal is another instance of complex order without appreciable information content." Michael Polanyi, "Life Transcending Physics and Chemistry," *Chemical and Engineering* **45**, 62 (1967). See also Marjorie Grene, ed., *Knowing and Being*, Essays by Michael Polanyi (Chicago: University of Chicago Press, 1968), p. 228.

131. Michael Polanyi, "Life Transcending Physics and Chemistry," p. 59.

132. Michael Polanyi, "Life Transcending Physics and Chemistry," and Grene, p. 229 and 232-233. Michael Polanyi held that the structure of machines and their workings are made by human beings and are not the random results of physical or chemical laws, even though their material and forces obey those laws. The design, shape, and operation of the machines are not due to physical and chemical forces and cannot be explained only by chemical and physical laws. Polanyi noted that the workings of a living organism have been compared to the working of machines with physiology interpreting the organism as a complex network of mechanisms. For example, the various organs of the body such as the stomach and liver function in a manner similar to a machine.

As a machine can never be reduced to the laws of physics and chemistry, so a living organism can never be reduced to these laws. In his words, "when I say that life transcends physics and chemistry, I mean that biology cannot explain life in our age by the current workings of physical and chemical laws." It is interesting to note that once Polanyi discovered his irreducible principles, he concluded that consciousness could never be reduced to physics and chemistry. But, again, once it is recognized, on other grounds, that life transcends physics and chemistry, there is no reason for suspending recognition of the obvious fact that consciousness is a principle that fundamentally transcends not only physics but also the mechanistic principles of living beings.

133. Grene, p. 241.

134. John Horgan, "From Complexity to Perplexity", *Scientific American* **272**, 106 (1995).

135. Horgan, "From Complexity to Perplexity", p. 106.

136. Behe, p. 179.

137. Horgan, "From Complexity to Perplexity", p. 107.

138. Naomi Oreskes, Kristin Shrader-Frechette, and Kenneth Belitz, "Verification, Validation, and Confirmation of Numerical Models in the Earth Sciences", *Science* **263**, 641 (1994).

139. Horgan, "From Complexity to Perplexity", p. 109.

140. Truth is the daughter of time.

141. This is also a fallacy in the reasoning of Richard Dawkins, another former Berkeley and current Oxford Professor, in Dawkins, *The Blind Watchmaker*, where he shows that a computer system can create complex patterns from simple algorithms.

142. See Stuart Kauffman, *At Home in the Universe: The Search for the Laws of Self-Organization and Complexity* (New York: Oxford University Press, 1995).

143. Richard Kerr, "Ancient Life on Mars", *Science* **273**, 864 (1996).

144. As quoted in W. W. Gibbs and C. S. Powell, "Bugs in the Data," *Scientific American* **275**, 20 (1996).

145. McKay, David S. et. al., "Search for Past Life on Mars: Possible Relic Biogenic Activity in Martian Meteorite ALH84001," *Science* **273**, 924 (1996).

146. Christopher F. Chyba, "Life Beyond Mars," *Nature* **382**, 576 (1996).

147. John Noble Wilford, "On Mars Life's Getting Tougher (If Not Impossible)", *New York Times*, December 22, 1996, pp. 1 and 6.

148. Monica Grady, Ian Wright and Colin Pillinger, "Opening a Martian Can of Worms?" *Nature* **382**, 575 (1996).

149. Kerr, p. 865.

150. "Fool's Gold on Mars?" *The Economist*, September 14, 1996, p. 83.

151. Grady, et. al., p. 575.

152. Kerr, p. 865.

153. Gibbs and Powell, p. 22.

154. Gibbs and Powell, p. 22.

155. Wilford, p. 6.

156. John P. Bradley, Ralph P. Harvey, and H.Y. McSween, Jr., "Magnetite Whiskers and Platelets in the ALH84001 Martian Meteorite: Evidence of Vapor Phase Growth," *Geochimica et Cosmochimica Acta* **60**, 5149 (1996).

157. Bradley, et. al., p. 5154.

158. Joseph L. Kirschvink, Altair T. Maine, and Hojatollah Vali, "Paleomagnetic Evidence of a Low-Temperature Origin of Carbonate in the Martian Meteorite ALH84001," *Science* **275**, 1629 (1997).

159. John W. Valley, John M. Eiler, Colin M. Graham, Everett K. Gibson, Christopher S. Romanek, and Edward M. Stopler, "Low-Temperature Carbonate Concretions in the Martian Meteorite ALH84001: Evidence from Stable Isotopes and Mineralogy," *Science* **275**, 1633 (1997).

160. John Noble Wilford, "Does Rock Show Past Life on Mars? The Jury is Still Out," *The New York Times*, p. A22, March 20, 1997.

161. Kathy Sawyer, "Digging into Data on Mars Life Claim," *The Washington Post*, p. A3, March 20, 1997. Sawyer reported that other teams detected biofilms which may indicate biological activity because they can be deposited by organic molecules by moving bacteria.

162. Harold J. Morowitz, "Past Life on Mars?" *Science* **273**, 1639 (1996).

163. Kerr, p. 866.

164. Gibbs and Powell, p. 22.

165. Arcady Mushegian and Eugene Koonin, "A Minimal Gene Set for Cellular Life Derived by Comparison of Complete Bacterial Genomes," *Proceedings of the National Academy of Sciences USA* **93**, 10268–10273 (1996).

166. Morowitz, "Past Life on Mars?", p. 1639.

167. Morowitz, *Beginnings of Cellular Life*, p. 12.

168. Grady, et. al., p. 576.

169. Ross, *The Creator and the Cosmos* (Colorado Springs, Colo.: NavPress Publishing Group, 1993), p. 155.

170. In Los Alamos in 1950 Enrico Fermi posed the following question to Edward Teller and two other physicists: If the universe is full of life, why haven't any aliens shown up on earth? There are so many stars older than our Sun and if life is common in the universe, aliens should have visited earth by now. The question has become known as Fermi's Paradox. E. Skindred writes about Fermi's query: "There are plenty of stars more ancient than our sun, Fermi noted, and if life is plentiful, it would have arisen on planets around these stars billions of years before it arose on earth. In that case, shouldn't earth have been visited or colonized by a race much older than our own? Even with relatively slow means of space travel, a civilization . . . could settle the galaxy in 5 million years or so. . . . It's the right question to ask,

scientists say, because it begins not with fantasy, but with a fact: We don't see aliens here now. To former NASA scientist Michael H. Hart, the absence of exotics on earth is compelling evidence that we are, if not the first, then among the first intelligent life forms to evolve anywhere in the galaxy. Both Harvard zoologist Ernst Mayr and astronomer Benjamin J. Zuckerman of the University of California, Los Angeles, second this position. The evolutionary path that leads to higher life is more complicated than others suppose, they hold. Since the beginning of life on earth, Mayr says, as many as 50 billion species have arisen, and only one of them has acquired technology. If intelligence has such high survival value, why don't we see more species develop it? " E. Skindred, "Where is Everybody?" *Science News* **150**, 153 (1996).

171. Yockey, "Self Organization of Life Scenarios and Information Theory," pp. 21 and 26.

172. Paul Davies, "The Unreasonable Effectiveness of Science", in John Marks Templeton, *Evidence of Purpose* (New York: The Continuum Publishing Company, 1996), p. 46.

173. This Doppler effect is given by the following equation:

$$\frac{\Delta\lambda}{\lambda_\circ} = \frac{v}{c}$$

where the wavelength shift $\Delta\lambda = \lambda - \lambda_\circ$ is computed with $\lambda_\circ$ representing the wavelength of a stationary source; λ represents the wavelength from a moving source; v is the velocity of the source along the line of sight between the observer and the source; and c is the speed of light. The speed of light is a constant at 299,792,458 km/s = 3×10^5 km/s = 183,262 miles per second. This is the speed of light in a vacuum. Light actually moves more slowly through glass, water, air or other transparent substances.

174. Mpc is the symbol for the term megaparsec. A megaparsec is one million parsecs. A parsec is a unit of length equal to 3.26 light years.

175. Wendy Freedman, "How Old is the Universe? New Measurements with the Hubble Space Telescope," *Science Spectra* **4**, 20 (1996). Subsequent to her article, Wendy Freedman and Gustav Tammann met at a conference in June, 1996, at Princeton Univer-

sity. Friedman announced her current estimate of the Hubble constant to be 70, give or take 10, and Tammann estimated 56, plus or minus 10, so that his high end overlaps her low end estimate.

176. Gustav A. Tammann and Allan Sandage, "The Hubble Diagram for Supernovae of Type Ia.II. The Effect on the Hubble Constant of a Correlation Between Absolute Magnitude and Light Decay Rate," *Astrophysical Journal* **452**, 16 (1995).

177. Joseph Silk, *A Short History of the Universe* (New York: Scientific American Library, 1994), p. 56.

178. The Schwarzschild radius is determined by the following equation:

$$R_{sch} = \frac{2GM}{c^2}$$

where c is the speed of light, G is the gravitational constant (6.668 x 10^{-11} Nm²kg⁻²).

Thus,

$$R_{sch} = 1.48 \times 10^{-27}M$$

Consequently, a black hole of 10 solar masses would have a Schwarzschild radius of 30km.

Inversely, with a Schwarzschild radius equal to one meter, the mass of a black hole will equal .67 x 10^{27}kg. This mass will be compressed into a sphere with a radius of one meter. The equation for the volume of a sphere is: $V = 4/3\pi r^3$. Mass divided by volume will yield the density of the matter within the black hole. (See Emiliani, p. 61.) One can verify that the density of the matter within the black hole exceeds the density of nuclear matter by the following calculation:

$$Density = .67 \times 10^{27}/4/3\pi r^3$$
$$= .16 \times 10^{27}kg/m^3$$
$$= .16 \times 10^{30}g/cm^3$$

179. Emiliani, pp. 61–62.

180. The distance (λc) of a wave or a Compton wavelength of an electron is given by the equation:

$$\lambda c = \frac{h}{mc}$$

where h is Planck's constant, m is the electron's mass, c is the speed of light. λ is the Compton wavelength.

181. The Planck mass may be expressed in the following equation:

$$m_p = \sqrt{\frac{hc}{G}} = 5.46 \times 10^{-8} \text{kg}$$

182. The Planck length is determined by the following equation:

$$\text{Planck length } (\lambda p) = \sqrt{\frac{Gh}{c^3}} = 4.05 \times 10^{-35} \text{m}$$

183. The Planck time is determined by the following equation:

$$(t_p) = \frac{\lambda c}{c} = \sqrt{\frac{Gh}{c^5}} = 1.35 \times 10^{-43} \text{s}$$

184. William Kaufmann III, *Universe*, 3rd edition (New York: W. H. Freeman and Company, 1991), p. 558.

185. As a White House Fellow working as an assistant to Vice President Nelson Rockefeller, one of my assignments was to meet from time to time with Dr. Edward Teller and convey his ideas to Vice President Rockefeller. His ideas were always creative and sometimes controversial. In the following poem originally published in *The New Yorker*, Dr. Harold P. Furth refers to a lecture by Dr. Edward Teller on the subject of antimatter. Antimatter and matter have a mirror image effect in space. Dr. Teller should actually greet Dr. Anti-Teller with his right hand while Dr. Anti-Teller should hold out his left hand. Of course the clasping of their hands caused their annihilation:

Perils of Modern Living
Harold P. Furth

Well up above the tropostrata
There is a region stark and stellar
Where, on a streak of anti-matter,
Lived Dr. Edward Anti-Teller.

Remote from Fusion's origin,
He lived unguessed and unaware
With all his antikith and kin,
and kept macassars on his chairs.

One morning, idling by the sea,
He spied a tin of monstrous girth
That bore three letters: A. E. C.
Out stepped a visitor from Earth.

Then, shouting gladly o'er the sands,
Met two who in their alien ways,
Were like as lentils. Their right hands
Clasped, and the rest was gamma rays.

(as quoted in Harold Fritzsch, *Quarks: The Stuff of Matter* (New York: Basic Books, Inc., 1983), pp. 28–29).

186. $E = 2(9.11 \times 10^{-31}$ kg$) (3.00 \times 10^8$ m/s$)^2$
  $= (1.64 \times 10^{-13}$J$) (1$ eV$/1.60 \times 10^{-19}$J$)$
  $= 1.02 \times 10^6$ eV or 1.02 MeV

187. Paul Zitewitz, Robert Neff and Mark Davids, *Physics* (Westerville, Ohio: Glencoe/McGraw-Hill, 1992), pp. 629–631.

188. Fritzsch, *Quarks*, p. 23.

189. Timothy Paul Smith, "Worlds Within Worlds," *The Sciences* **36**, 28 (1996).

190. Gordon Kane, *The Particle Garden* (New York: Addison-Wesley Publishing Company, Inc., 1995), p. 56.

191. Kane, p. 58.

192. Alan Isaacs, ed., *A Dictionary of Physics*, 3rd edition (New York: Oxford University Press, 1996), pp. 126–127.

193. Temperature, energy, radiation frequency and matter are related in the following equation:

$$e = mc^2 = h\upsilon = 3/2\, kT$$

where e=energy, m=mass, c=speed of light or 2.998×10^8m/s, h=Planck's constant or 6.626×10^{-34}Js, υ=frequency, k=Boltzmann constant or 1.380×10^{-23}Jk^{-1}, T = absolute temperature. If the mass of a particle is known, the corresponding amount of energy can be calculated as well as the temperature. The critical temperature is defined as the temperature above which a particle becomes unstable so that an increase in pressure cannot make a gas liquid. The critical temperature of a particle is directly related to its mass. This is the temperature at which mass and energy may be interchangeable. See Emiliani, p. 83.

194. Emiliani, p. 83.

195. Kaufmann, p. 588.

196. Robert Matthews, *Unraveling the Mind of God* (London: Virgin Publishing Ltd., 1993), p. 205.

197. Matthews, p. 205.

198. Matthews, p. 211.

199. Hugh Ross, *Beyond the Cosmos* (Colorado Springs, Colo.: NavPress Publishing Group, 1996), p. 30.

200. Here and everywhere.

201. Ross, *Beyond the Cosmos*, p. 32.

202. Gary Taubes, "How Black Holes May Get String Theory Out of a Bind," *Science* **268**, 1699 (1995).

203. Gary Taubes, "A Theory of Everything Takes Shape," *Science* **269**, 1513 (1995).

204. See Stephen Hawking and George Ellis, "The Cosmic Blackbody Radiation and the Existence of Singularities in Our Universe," *Astrophysical Journal* **152**, 25 (1968).

205. This lengthening of the wavelength follows the law named after German physicist Wilhelm Wien, who discovered the relationship between the dominant wavelength (λmax) of the energy (measured in meters) from a blackbody and its temperature (measured in Kelvin degrees):

$$\lambda \, max = \frac{2.9 \times 10^{-3}}{T}$$

The higher the temperature of an object, the shorter the dominant wavelength of its electromagnetic energy.

206. John Leslie, *Universes* (London: Routledge, 1989), p. 52.

207. John Gribbin, Bernard Carr, and Martin Rees, *Cosmic Coincidences* (New York: Bantam, 1989).

208. Ross, *The Creator and the Cosmos*.

209. Fred Hereen, *Show Me God* (Wheeling, Il.: Search Light Publishing, 1995).

210. For a more thorough discussion of the triple alpha process, see Silk, pp. 18–19.

211. Owen Gingerich, "Kepler's Anguish and Hawking's Queries: Reflections on Natural Theology," *The Great Ideas Today* (Chicago: Encyclopedia Britannica, Inc., 1992), pp. 273–274.

212. Fred Hoyle as quoted in Heeren, p. 179.

213. Alan Guth, "Inflationary Universe: A Possible Solution to the Horizon and Flatness Problems", *Physical Review D* **23**, 348 (1981).

214. Bernard Lovell, *In the Center of Immensities* (New York: Harper & Row, 1978), pp. 122–123.

215. Paul Davies, *God and the New Physics* (London: J. M. Dent and Sons, Ltd., 1983), p. 179.

216. The critical density is determined by the following equation:

$$P_c = \frac{3H_o^{\ 2}}{8\pi G}$$

where H_o is the Hubble constant and G is the gravitational constant (6.668×10^{-11} Nm^{2}kg^{-2}). If we assume a Hubble constant of 50 km/s Mpc, $P_c = 4.8 \times 10^{-27}$ kg/m^3. At this value for P_c we have approximately three hydrogen atoms for every cubic meter of space.

217. Kaufmann, p. 582.

218. Leslie, p. 6.

219. Paul Davies, *The Accidental Universe* (Cambridge: Cambridge University Press, 1982), pp. 70–71.

220. John Polkinghorne, "A Potent Universe," in Templeton, p. 111.

221. Davies, *The Accidental Universe*, p. 73.

222. Leslie, p. 6.

223. Ross, *The Creator and the Cosmos*, p. 115.

224. Hugh Ross, *The Fingerprint of God*, 2nd edition (Orange, Calif.: Promise Publishing, 1991), p. 123.

225. Stephen W. Hawking, *A Brief History of Time: From the Big Bang to Black Holes* (New York: Bantam Books, 1988), p. 125.

226. Polkinghorne, *Beyond Science* (Cambridge: Cambridge University Press, 1996), p. 85.

227. Leslie, p. 40.

228. Davies, *God and the New Physics*, pp. 178–179.

229. Penrose, p. 344.

230. Davies, *The Accidental Universe*, op. cit., pp. 60–61.

231. Davies, p. 68.

232. Hawking, *A Brief History of Time*, pp. 163–165.

233. John Wheeler as cited in Leslie, p. 47.

234. Leslie, p. 43.

235. Leslie, p. 43.

236. A language well made or suited.

237. J. H. Taylor, et. al., "Experimental Constraints on Strong-Field Relativistic Gravity," *Nature* **355**, 132 (1992).

238. Pearcey and Thaxton, p. 173.

239. Paul Davies, *The Mind of God* (New York: Simon and Schuster, 1992), p. 151.

240. Paul Davies, "The Unreasonable Effectiveness of Science," in Templeton, p. 554.

241. Edward Harrison, *Masks of the Universe* (New York: Macmillan Publishing Company, 1985), p. 164. See Kenneth S. Krane, *Modern Physics* (New York: John Wiley & Sons, Inc. 1996), pp. 492–494.

242. Kaufmann, p. 554.

243. R.F.C. Vessot et. al., "Test of Relativistic Gravitation and Space-Borne Hydrogen Maser," *Physical Review Letters* **45**, 2081 (1980).

244. Hawking, *A Brief History of Time*, p. 52. Another confirmation of Einstein's theory of general relativity was the verification of a prediction from the equations that the tidal forces in a strong gravitational field would cause gravitational waves to radiate from a binary pair of neutron stars. The distance between these stars was initially one million kilometers. According to the prediction, the stars should spiral towards each other a few centimeters each year. The spiraling stars function as extremely accurate clocks with their orbits timed with a remarkable precision. The prediction was verified by observation to an amazing accuracy of 99.7 percent. See Silk, p. 145.

245. Davies, *The Mind of God*, p. 152.

246. Polkinghorne, *Beyond Science*, pp. 79–80.

247. Leslie, pp. 64–65.

248. William K. Hartmann and Chris Impey, *Astronomy*, 5th edition (Belmont, Calif.: Wadsworth Publishing Company, 1994), p. 640.

249. The formula is:

$$S_{bh} = \frac{A}{4} \times \frac{kc^3}{Gh_p}$$

where A is the black hole's horizon surface area, k is the Boltzmann constant (1.380×10^{-23} Jk^{-1}), c is the speed of light (2.998×10^8 m/s), G is the gravitational constant (6.668×10^{-11} Nm^{2}kg^{-2}), h_p is Planck's constant (6.626×10^{-34}Js) over 2π.

250. The relation is as follows:

$$A = m^2 \times 8\pi \, (G^2/c^4)$$

Substituting in the immediately preceding equation we have:

$$S_{bh} = m^2 \times 2\pi \, (kG/h_p c)$$

251. Penrose, pp. 339–343.

252. Lovell, p. 103.

253. Lovell, p. 103.

254. John D. Barrow, *The Origin of the Universe* (New York: Basic Books, 1994), p. 90. Joseph Silk and John D. Barrow describe Planck time in the following terms:

In a universe governed partially by the influence of gravity, light propagation, and a quantum theory of matter, there exists a unique time at which all these effects are of equal importance. The strength of gravity is described by Newton's constant of nature, $G = 6.672 \times 10^{-8}$ cm^3 gm^{-1} sec^{-2}, the uncertainty of quantum theory via Planck's constant, $h = 6.625 \times 10^{-27}$ gm cm^2 sec^{-1}, and the relativistic theory of light characterized by the speed of light, $c = 3.0 \times 10^{10}$ cm sec^{-1}. If we look at the units of mass (gm), length (cm) and time (sec) we have chosen, we see that it is possible to combine these three constants of nature and create a mixture with the units of a time in one and only one way. Take G, multiply it by h, then divide by c five times, now take the square root of your answer; notice that the units of your answer are seconds and its magnitude, t_p, is

$$t_p = \sqrt{\frac{Gh}{c^5}} = 1.33 \times 10^{-43} \text{ sec}$$

This is undoubtedly the shortest interval of time the reader has ever encountered. To put it in perspective, the time it takes for light to cross an atomic nucleus, 10^{-24} second, is huge by comparison. The time t_p, although first discovered by the Irish physicist, George Johnstone-Stoney, in the 1870s, is called the Planck time after Max Planck, one of the pioneers of quantum theory who discovered it independently in 1906. John D. Barrow and Joseph Silk, *The Left Hand of Creation* (Oxford: Oxford University Press, 1993), pp. 61–62.

255. Robert Jastrow, *God and the Astronomers* (New York: Harper & Row, 1979), p. 250.

256. The Compton wavelength for a particle at rest is h/mc, where h is the Planck constant and c is the speed of light. As indi-

cated above, this is the smallest length scale in quantum physics. The Schwarzchild radius gives a mass of $(hc/2G)^{1/2}$ or the Planck mass (mp) which is equal to 10^{-5} gram. The classical theory of gravity stops at an elementary particle of this mass. This is the smallest scale and highest density to which such a classical theory can apply. The laws of physics begin at this scale which can also be represented by an energy scale of 10^{19} GeV (GeV means gigaelectron volt equal to 10^9ev or BeV, billion-electron volt), a length of scale of 10^{-33} centimeters, a time scale of 10^{-43} seconds or Planck time. At Planck time, the density of matter is a remarkable 10^{94} grams per cubic centimeter. See Silk, p. 75.

257. Kaufmann, p. 578.

258. Richard Morris, *Cosmic Questions* (New York: John Wiley and Sons, Inc., 1993), p. 140.

259. Ross, *The Creator and the Cosmos*, p. 97.

260. Ross, *The Creator and the Cosmos*, p. 97.

261. Personal correspondence from John Polkinghorne, dated October 31, 1996.

262. Nothing comes from nothing.

263. Keith Ward, *God, Chance & Necessity*, Oxford: One World (1996), p. 40.

264. Ward, pp. 39–40.

265. M. A. Corey, *God and the New Cosmology* (Lanham, Md.: Rowman and Littlefield, 1993), p. 43.

266. Heinz Pagels, *Perfect Symmetry* (New York: Bantam Books, 1986), p. 365.

267. Albert Einstein, *Ideas and Opinions—The World As I See It* (New York: Bonanza Books, 1931), p. 40.

268. Freeman Dyson, *Disturbing the Universe* (New York: Harper & Row, 1979), p. 250.

269. To know truly is to know causes. Thinking about a logic or intelligence that exists prior to space and time is obviously difficult. As we have noted, we should not even use the word "prior"

because time begins at Planck time and cannot be part of the concept of "true nothingness." If time and space came into existence with the Big Bang, the conclusion becomes invalid that the beginning of the universe in time would have been preceded by a time. But this implies that the *very initial* Big Bang itself was not a temporal event. This conclusion has profound implications for physical, metaphysical, and theological formulations. German theologian Wolfhart Pannenberg, who physicist Frank Tipler has credited with increasing his understanding of many physical concepts, has written: "This suggests to theology a new formulation of the idea of creation: the divine act of creation does not occur in time—rather, it constitutes an eternal act, contemporaneous with all time, that is, with the entire world process. Yet this world process itself has a temporal beginning, because it takes place in time. In this sentence, I assert that eternity itself is described by statements of time. With a musical parable one might speak of eternity as the sounding together of all time in a sole present. Elsewhere I have developed this concept of eternity from the human experience of time, from the relativity of the distinction of past, present, and future corresponding to the relativity of the directions in space. In view of the relativity of the modes of time to the aspect of the human being experiencing time, this resulted in the assumption that all time, if it could be, so to speak, surveyed from a 'place' outside the course of time, would have to appear as contemporaneous. . . . Understood in the sense of the suggestions above, the concept of eternity comprehends all time and everything temporal in itself—a conception of the relationship of time and eternity that goes back to Augustine and is connected to the Israelite understanding of eternity as unlimited duration throughout time. The world view of the theory of relativity also can be understood in the sense of a last contemporaneousness of all events that for us are partitioned into a temporal sequence." Wolfhart Pannenberg, *Toward a Theology of Nature* (Louisville: Westminster/John Knox Press, 1993), pp. 100–101. Remember Einstein's statement that the distinction between past, present, and future is in many respects a stubborn illusion? These are difficult concepts considering our personal sensations of the passage of time, but they fit well with the metaphysical discussions contained in this book.

270. Stephen Hawking and Roger Penrose, "The Singularities of Gravitational Collapse and Cosmology," *Proceedings of the Royal Society of London Series A*, **314**, 529–548 (1970).

271. Roy E. Peacock, *A Brief History of Eternity*, (Wheaton, Il.: Crossway Books, 1990), p. 24.

272. It lacks a beginning and an end.

273. James B. Hartle and Stephen W. Hawking, "Wave Function of the Universe," *Physical Review D* **28**, 2974, 2975 (1983).

274. More in possibility than in fact.

275. Imaginary numbers have unusual characteristics. For example, the square of an imaginary number can be a negative number.

276. Hawking, *A Brief History in Time*, pp. 136, 137 and 139.

277. According to truth.

278. Ward, pp. 41, 43 and 44.

279. Kitty Ferguson, *The Fire in the Equations* (Grand Rapids: W. Eerdmans Publishing Company, Inc., 1994), p. 137.

280. Ferguson, pp. 113 and 128.

281. Ferguson, p. 139.

282. For the sake of argument.

283. Ward, pp. 23–24.

284. For a complete discussion of this *causa essendi* argument see Mortimer Adler, *How to Think About God: A Guide for the 20th Century Pagan*. (New York: MacMillan Publishing Co., Inc., 1980).

285. Gingerich (1992), p. 284.

286. Peacock, p. 95.

287. Morris, p. 152.

288. Murray Gell-Mann, *The Quark and the Jaguar* (New York: W. H. Freeman and Company, 1994), pp. 220–221.

289. Hawking, *A Brief History of Time*, pp. 149–150.

290. Huw Price, "A Point on the Arrow of Time", *Nature* **340**, 181 (1989).

291. Hawking, *A Brief History of Time*, pp. 147–148.

292. Price, "A Point on the Arrow of Time", p. 181.

293. Price, "A Point on the Arrow of Time," p. 182.

294. Huw Price, *Time's Arrow and Archimedes Point* (New York: Oxford University Press, 1996), p. 92.

295. Peacock, p. 114.

296. Silk, p. 9.

297. Owen Gingerich, "Dare a Scientist Believe in Design?" in Templeton, p. 26.

298. Polkinghorne, *Beyond Science*, pp. 87–88.

299. Alan Guth as quoted in Hereen, p. 311.

300. Pagels, *Perfect Symmetry*, pp. 377–378.

301. Gribbin, p. xiii.

302. Hereen, p. 224.

303. Polkinghorne, in Templeton, p. 114.

304. Friedrich Nietzche, *The Joyful Wisdom*, trans. by Thomas Common (New York: Russell and Russell, Inc., 1964), p. 25.

305. Someone who knows right from wrong.

306. Francis A. Schaeffer, *How Should We Then Live?* (Old Tappan, N.J.: Fleming H. Revell Co., 1976), p. 122.

307. I believe even though it is absurd.

SELECTED BIBLIOGRAPHY

BOOKS

Abbott, Edwin A. *Flatland*. New York: Signet, 1984.

Adair, Robert. *The Great Design: Particles, Fields and Creation*. Oxford: Oxford University Press, 1987.

Adler, Mortimer. *How to Think About God: A Guide for the 20th Century Pagan*. New York: Macmillan, 1980.

Auyang, Sunny. *How is Quantum Field Theory Possible?* Oxford: Oxford Unversity Press, 1995.

Barbour, Ian. *Issues in Science and Religion*. New York: Harper & Row, 1971.

Barrow, John, D. *The Origin of the Universe*. New York: Basic Books, 1994.

————, and Joseph Silk. *The Left Hand of Creation: The Origin and Evolution of the Universe*. 2nd ed. New York: Oxford University Press, 1994.

————, and Frank Tipler. *The Anthropic Cosmological Principle*, Oxford: Oxford University Press, 1986.

Behe, Michael J. *Darwin's Black Box*. New York: The Free Press, 1996.

Benjamin, Arthur, and Michael Brant Shermer. *Mathemagics*. Los Angeles: Lowell House, 1993.

Brooks, J., and G. Shaw. *Origin and Development of Living Systems*. New York: Academic Press, 1973.

Cairns-Smith, A.G. *Genetic Takeover and the Mineral Origins of Life*. Cambridge: Cambridge University Press, 1982.

————. *The Life Puzzle*. Edinburgh: Oliver and Boyd, 1971.

Carroll, Lewis. *Alice in Wonderland*. New York: The New American Library, 1960.

————. *Through the Looking Glass*. New York: The New American Library, 1960.

Clark, Ronald. *Einstein: The Life and Times*. New York: World Publishing, 1971.

Close, F., M. Marten, and C. Sutton. *The Particle Explosion*. New York: Oxford University Press, 1987.

Corey, M.A. *God and the New Cosmology*. Lanham: Rowman and Littlefield, 1993.

Cornell, James, ed. *Bubbles, Voids and Bumps in Time: The New Cosmology*. Cambridge: Cambridge University Press, 1989.

Coughlin, G.D., and J.E. Dodd. *The Ideas of Particle Physics*. 2nd ed. Cambridge: Cambridge University Press, 1991.

Craig, Harding, ed. *William Shakespeare, King Henry IV, The Complete Works of Shakespeare*. Chicago: Scott Foresman, 1961.

Crick, F. *Life Itself*. New York: Simon and Schuster, 1981.

Davies, Paul. *The Accidental Universe*. Cambridge: Cambridge University Press, 1982.

———. *The Cosmic Blueprint*. New York: Simon and Schuster, 1988.

———. *God and the New Physics*. London: J.M. Dent, 1983.

———. *The Mind of God*. New York: Simon and Schuster, 1992.

———. *Superforce*. New York: Simon and Schuster, 1984.

———. (ed.) *The New Physics*. Cambridge: Cambridge University Press, 1989.

———, and Julian Brown. *Superstrings: A Theory of Everything?* Cambridge: Cambridge University Press, 1988.

Dawkins, Richard. *The Blind Watchmaker*. New York: W.W. Norton, 1986.

Delbrück, Max. *Mind From Matter?* Oxford: Blackwell Scientific Publications, 1986.

Denton, Michael. *Evolution: A Theory in Crisis*. Bethesda: Adler and Adler, 1985.

Dyson, Freeman. *Disturbing The Universe*. New York: Harper & Row, 1979.

Einstein, Albert. *The Meaning of Relativity*, 5th ed. New York: MJF, 1984.

———. *Relativity: The Special and the General Theory*. New York: Crown Trade Paperbacks, 1961.

———. *Ideas and Opinions—The World As I See It*. New York: Bonanza, 1931.

Eisberg, R., and R. Resnick. *Quantum Physics of Atoms, Molecules, Solids, Nuclei, and Particles*. 2nd ed. New York: John Wiley, 1985.

Emiliani, Cesare. *The Scientific Companion*. New York: John Wiley, 1995.

Engel, S. Morris. *With Good Reason*. New York: St. Martin's Press, 1976.

Faletta, Nicholas. *The Paradoxicon*. New York: John Wiley, 1990.

Ferguson, Kitty. *The Fire in the Equations*. Grand Rapids: W. Eerdmans, 1994.

Fritzsch, Harold. *Quarks: The Stuff of Matter*. New York: Basic, 1983.

Gardner, Martin. *HexaFlexegons and Other Mathematical Diversions*. Chicago: The University of Chicago Press, 1988.

Gell-Mann, Murray. *The Quark and the Jaguar*. New York: W.H. Freeman, 1994.

Gingerich, Owen. *Kepler's Anguish and Hawking's Queries: Reflections on Natural Theology, The Great Ideas Today*. Chicago: Encyclopedia Britannica, 1992.

Gleick, James. *Chaos: Making a New Science*. New York: Penguin, 1987.

Grene, Marjorie, ed. *Knowing and Being*, Essays by Michael Polanyi. Chicago: University of Chicago Press, 1968.

Gribbin, John. *In the Beginning: After COBE and Before the Big Bang*. New York: Little, Brown and Company, 1993.

———. *In Search of Schrödinger's Cat*. New York: Bantam, 1984..

———, Bernard Carr, and Martin Rees. *Cosmic Coincidences*. New York: Bantam, 1989.

Haldane, J.B.S. *The Origin of Life*. In *Rationalist Annvel*. London: C.A. Watts & Co., 1929.

Harrison, Edward. *Cosmology*. Cambridge: Cambridge University Press, 1981.

———. *Masks of the Universe*. New York: Macmillan, 1985.

Hartmann, William K., and Chris Impey. *Astronomy*. 5th ed. Belmont, Calif.: Wadsworth, 1994.

Hawking, Stephen. *A Brief History of Time*. New York: Bantam, 1988.

———, and Roger Penrose. *The Nature of Space and Time*. Princeton: Princeton University Press, 1996.

Hereen, Fred. *Show Me God*. Wheeling, Il.: Search Light, 1995.

Heller, Joseph. *Catch-22*. New York: Bantam Doubleday Dell, 1955.

Hilgevoord, Jan, ed. *Physics and Our View of the World*. Cambridge: Cambridge University Press, 1994.

Horgan, John. *The End of Science*. Reading, Pa.: Addison-Wesley, 1995.

Hoyle, Fred. *The Intelligent Universe*. London: Michael Joseph, 1983.

———, and Chandra Wickramasinghe. *Evolution from Space*. London: J.M. Dent, 1981.

———, and Chandra Wickramasinghe. *Our Place in the Cosmos*. London: J.M. Dent, 1993.

Jastrow, Robert. *God and the Astronomers*. New York: Harper & Row, 1979.

Johnson, George. *Fire in the Mind*. New York: Knopf, 1995.

Johnson, Phillip E. *Darwin on Trial*. Washington, D.C.: Regnery Gateway, 1991.

LeComte du Noüy, Pierre. *Human Destiny*. New York: Longmans, Green and Co., 1947.

Kaku, Michio. *Beyond Einstein*. New York: Anchor, 1995.

Kane, Gordon. *The Particle Garden*. New York: Addison-Wesley, 1995.

Kauffman, Stuart. *At Home in the Universe: The Search for the Laws of Self-Organization and Complexity*. New York: Oxford University Press, 1995.

———. *The Origins of Order*. New York: Oxford University Press, 1993.

Kaufmann III, William. *Universe*. 3rd ed. New York: W.H. Freeman, 1991.

Krane, Kenneth. *Modern Physics*. 2nd ed. New York: John Wiley, 1996.

Küppers, Bernd-Olaf. *Information and the Origin of Life*. Cambridge, Mass.: The MIT Press, 1990.

Lewin, Roger. *Complexity*. New York: Macmillan, 1992.

Leslie, John. *Universes*. London: Routledge, 1989.

———. Leslie, John (ed.). *Physical Cosmology of Philosophy*. New York: Macmillan, 1990.

Lovell, Bernard. *In the Center of Immensities*. New York: Harper & Row, 1978.

Matthews, Robert. *Unraveling the Mind of God*. London: Virgin, 1993.

Milani, Jean P. et al. *Biological Science*. Lexington, Mass.: D.C. Heath and Company, 1990.

Miller, Stanley, and Orgel. *The Origins of Life on Earth*. Englewood Cliffs, N.J.: Prentice-Hall, 1974.

Monod, Jacques. *Chance and Necessity*. London: Collins, 1972.

Moorhead, Paul S. and Martin M. Kaplan, (eds.). *Mathematical Challenges to the Neo-Darwinian Interpretation of Evolution*. Philadelphia: The Wister Institute Press, 1997.

Moreland, J.P., ed. *The Creation Hypothesis*. Downers Grove, Il.: InterVarsity Press, 1994.

Morowitz, Harold J. *Beginnings of Cellular Life*. New Haven: Yale University Press, 1992.

———. *Energy Flow in Biology*. Woodbridge, Conn.: Ox Bow Press, 1979.

Morris, Richard. *Cosmic Questions*. New York: John Wiley, 1993.

Nagel, Ernst, and James Newman. *Gödel's Proof*. New York: New York University Press, 1986.

Nelson, Edward. *Quantum Fluctuations*. Princeton: Princeton University Press, 1985.

Nichols, G., and Prigogine I. *Self-Organization in Nonequilibrium Systems*. New York: John Wiley, 1977

Nietzsche, Friedrich. *The Joyful Wisdom*, trans. by Thomas Common. New York: Russell and Russell, 1964.

Ninio, Jacques. *Molecular Approaches to Evolution*. Princeton: Princeton University Press, 1993.

Ogilvy, C. Stanley. *Excursions in Mathematics*. Mineola, New York: Dover, 1994.

Olivastro, Dominic. *Ancient Puzzles*, New York: Bantam, 1993.

Oparin, A.I. Proiskhozdenic Zhizny. Moscow: Izd Moskovski Rabochi, 1924. Published in English translation in J.D. Bernad. *The Origin of Life*. London: Weidenfield and Nicolson, 1967.

Pagels, Heinz. *Perfect Symmetry*. New York: Bantam, 1986.

Pannenberg, Wolfhart. *Toward a Theology of Nature*. Louisville: Westminster/John Knox Press, 1993.

Parks, Howard, Gary Musser, Robert Burton, and William Siebler. *Mathematics in Life, Society & the World*. Upper Saddle River: N.J.: Prentice-Hall, 1997.

Peacock, Roy E. *A Brief History of Eternity*. Wheaton: Crossway, 1990.

Peacocke, Arthur. *God and the New Biology*. Gloucester, Mass.: Peter Smith, 1994.

Pearcey, Nancy R., and Charles B. Thaxton. *The Soul of Science*. Wheaton, Il.: Crossway, 1994.

Peat, F. *Superstrings and the Search for the Theory of Everything*. New York: Contemporary, 1988.

Peebles, P.J.E. *Principles of Physical Cosmology*. Princeton: Princeton University Press, 1993.

Penrose, Roger. *The Emperor's New Mind*. New York: Oxford University Press, 1989.

Polya, G. *How to Solve It: A New Aspect of Mathematical Method*. 2nd ed. Princeton: Princeton University Press, 1988.

Polanyi, Michael. *Personal Knowledge*. New York: Harper & Row, 1958.

Polkinghorne, John. *Beyond Science*. Cambridge: Cambridge University Press, 1996.

———. *Science and Creation*. London: SPCK, 1988.

Price, Huw. *Time's Arrow and Archimedes Point*. New York: Oxford University Press, 1996.

Prigogine, I. *From Being to Becoming*. San Francisco: W.H. Freeman and Co., 1980.

———, and I. Stengers. *Order Out of Chaos*. London: Heinemann, 1984.

Ross, Hugh. *Beyond the Cosmos*. Colorado Springs: NavPress, 1996.

———. *The Creator and the Cosmos*. Colorado Springs: NavPress, 1993.

———. *The Fingerprint of God*. 2nd ed. Orange, Calif.: Promise, 1991.

Rucker, Rudy. *Infinity and the Mind*. New York: Bantam, 1983.

Ruelle, David. *Chance and Chaos*. Princeton: Princeton University Press, 1991.

Schaeffer, Francis A. *How Should We Then Live?* Old Tappan, N.J.: Fleming H. Revell, 1976.

Schrodinger, E. *What is Life?* Cambridge: Cambridge University Press, 1944.

Schroeder, Gerald L. *Genesis and the Big Bang*. New York: Bantam, 1990.

Shapiro, Robert. *Origins: A Skeptic's Guide to the Creation of Life on Earth*. New York: Simon and Schuster, Summit, 1986.

Silk, Joseph. *A Short History of the Universe*. New York: Scientific American Library, 1994.

Templeton, John, ed. *Evidence of Purpose*. New York: Continuum, 1996.

Thaxton, Charles B., Walter L. Bradley, and Roger Olsen. *The Mystery of Life's Origin: Reassessing Current Theories*. New York: Philosophical Library, 1984.

Tipler, Frank. *The Physics of Immortality*. New York: Doubleday, 1994.

Twain, Mark. *Life on the Mississippi*. New York: Bantam, 1981.

Yang, C.N. *Elementary Particles*. Princeton: Princeton University Press, 1961.

Yockey, Hubert. *Information Theory and Molecular Biology*. Cambridge: Cambridge University Press, 1992.

Ward, Keith. *God, Chance & Necessity*. Oxford: One World, 1996.

Weinberg, Steven. *Dreams of a Final Theory*. New York: Pantheon, 1992.

———. *The First Three Minutes*. New York: Bantam, 1984.

Zee, A. *Fearful Symmetry: The Search for Beauty in Modern Physics*. New York: Macmillan, 1986.

Zitewitz, Paul, Robert Neff, and Mark Davids. *Physics*. Westerville, Ohio: Glencoe/McGraw-Hill, 1992.

ARTICLES, PERIODICALS AND PROCEEDINGS

Bludman, S.A. "Thermodynamics and the End of a Closed Universe," *Nature* **308**, 319 (1984).

Barrow, J.D., and J. Silk. "The Structure of the Early Universe," *Scientific American* **242**, 118 (1980).

Bradley, J.P., R.P. Harvey, and H.Y. McSween, Jr. "Magnetite Whiskers and Platelets in the ALH84001 Martian Meteorite: Evidence of Vapor Phase Growth," *Geochimica et Cosmochimica Acta* **60**, 5149 (1996).

Carver, J.H. "Prebiotic Atmospheric Oxygen Levels," *Nature* **292**, 136 (1981).

Chyba, Christopher F. "Life Beyond Mars," *Nature* **382**, 576 (1996).

Fraser, Gordon. "The Human Paradox: Stephen Hawking and His Work," *Science Spectra* **4**, 13 (1996).

Freedman, Wendy. "The Expansion Rate and Size of the Universe," *Scientific American* **267**, 54 (1992).

———. "How Old is the Universe? New Measurements with the Hubble Space Telescope," *Science Spectra* **4**, 20 (1996).

Gibbs, W.W., and C.S. Powell. "Bugs in the Data," *Scientific American* **275**, 20 (1996).

Glashow, S.L. "Quarks with Color and Flavor," *Scientific American* **233**, 38 (1975).

Grady, Monica, Ian Wright, and Colin Pillinger. "Opening a Martian Can of Worms?" *Nature* **382**, 575 (1996).

Guth, Alan. "Inflationary Universe: A Possible Solution to the Horizon and Flatness Problems," *Physical Review D* **23**, 348 (1981).

———. "The Impossibility of a Bouncing Universe" *Nature* **302**, 505 (1983).

———, and Paul Steinhardt, "The Inflationary Universe," *Scientific American*, May 1984, pp.116-120.

Hart, Michael. "The Evolution of the Atmosphere of the Earth," *Icarus* **33**, 23 (1978).

Hartle, James B., and Stephen W. Hawking. "Wave Function of the Universe," *Physical Review D* **28**, 2960 (1983).

Hawking, Stephen, and George Ellis. "The Cosmic Blackbody Radiation and the Existence of Singularities in Our Universe," *Astrophysical Journal* **152**, 25 (1968).

———, and Roger Penrose. "The Singularities of Gravitational Collapse and Cosmology," *Proceedings of the Royal Society of London, Series A* **314**, 529 (1970).

Hayes, John M. "The Earliest Memories of Life on Earth," *Nature* **384**, 21 (1996).

Horgan, John. "From Complexity to Perplexity," *Scientific American* **272**, 106 (1995).

———. "In the Beginning," *Scientific American* **264**, 121 (1991).

Huber, Claudia, and Günter Wächtershäuser. "Activated Acetic Acid by Carbon Fixation on (Fe, Ni)s Under Primordial Conditions," *Science* **276**, 245 (1997).

Kerr, Richard. "Ancient Life on Mars," *Science* **273**, 864 (1996).

Kirschvink, Joseph L., Altair T. Maine, and Hojatollah Vali, "Paleomagnetic Evidence of a Low-Temperature Origin of Carbonate in the Martian Meteorite ALH84001," *Science* **275**, 1629 (1997).

Maddox, John. "The Anthropic View of Nucleosynthesis," *Nature* **355**, 107 (1992).

McKay, David S. et al. "Search for Past Life on Mars: Possible Relic Biogenic Activity in Martian Meteorite ALH84001," *Science* **273**, 924 (1996).

Miller, Stanley L. "A Production of Amino Acids Under Possible Primitive Earth Conditions," *Science* **117**, 528 (1953).

Mojzsis, S.J. et. al. "Evidence for Life on Earth Before 3,800 Million Years Ago," *Nature* **384**, 55 (1996).

Morowitz, Harold J. "Past Life on Mars?" *Science* **273**, 1639 (1996).

Mushegian, Arcady, and Eugene Koonin. "A Minimal Gene Set for Cellular Life Derived by Comparison of Complete Bacterial Genomes," *Proceedings of the National Academy of Sciences USA*, **93**, 10268 (1996).

Oreskes, Naomi, Kristin Shrader-Frechette, and Kenneth Belitz. "Verification, Validation, and Confirmation of Numerical Models in the Earth Sciences," *Science* **263**, 641 (1994).

Orgel, Leslie. "RNA Catalysis and the Origin of Life," *Journal of Theoretical Biology* **123**, 127 (1986).

Polanyi, Michael. "Life Transcending Physics and Chemistry," *Chemical and Engineering* **45**, 62 (1967).

Price, Huw. "A Point on the Arrow of Time," *Nature* **340**, 181 (1989).

Quigg, C. "Elementary Particles and Forces," *Scientific American* **255**, 84 (1985).

Sawyer, Kathy. "Digging into Data on Mars Life Claim," *The Washington Post*, p.A3, March 20, 1997.

Schidlowski, Manfred. "A 3,800 million year Isotopic Record of Life from Carbon in Sedimentary Rocks," *Nature* **333**, 313 (1988).

Shapiro, Robert. "Prebiotic Ribose Synthesis: A Critical Analysis," *Origins of Life and Evolution of the Biosphere* **18** (1988).

Simpson, George Gaylord. "The Nonprevalence of Humanoids," *Science* **143**, 771 (1964).

Skindred, E. "Where is Everybody?" *Science News*, **150**, 153 (1996).

Smith, Timothy Paul. "Worlds Within Worlds," *The Sciences* **36**, 28 (1996).

Stewart, Ian. "The Interrogator's Fallacy," *Scientific American* **275**, 172 (1996).

Tammann, Gustav A., and Allan Sandage. "The Hubble Diagram for Supernovae of Type Ia.II. The Effect on the Hubble Constant of a Correlation Between Absolute Magnitude and Light Decay Rate," *Astrophysical Journal* **452**, 16 (1995).

Taubes, Gary."A Theory of Everything Takes Shape," *Science* **269**, 1513 (1995).

———. "How Black Holes May Get String Theory Out of a Bind," *Science* **268**, 1699 (1995).

Taylor, J.H. et. al. "Experimental Constraints on Strong-Field Relativistic Gravity," *Nature* **355**, 132 (1992).

Valley, John W., John M. Eiler, Colin M. Graham, Everett K. Gibson, Christopher S. Romanek, and Edward M. Stopler, "Low-Temperature Carbonate Concretions in the Martian Meteorite ALH84001: Evidence from Stable Isotopes and Mineralogy," *Science* **275**, 1633 (1997).

Vessot, R.F.C. et. al. "Test of Relativistic Gravitation and Space-Borne Hydrogen Maser," *Physical Review Letters* **45**, 2081 (1980).

Wald, George. "The Origin of Life," *Scientific American* **191**, 48 (1954).

Wilford, John Noble. "Does Rock Show Past Life on Mars? The Jury Is Still Out," *New York Times*, March 20, 1997, p. A22.

——— "On Mars Life's Getting Tougher (If Not Impossible)" *New York Times*, December 22, 1996, pp. 1 and 6.

Yockey, Hubert. "Comments on Let There Be Life: Thermodynamic Reflections on Biogenesis and Evolution by Avshalom C. Elitzur," *Journal of Theoretical Biology* **176**, 349 (1995).

———. "A Calculation of the Probability of Spontaneous Biogenesis by Information Theory," *Journal of Theoretical Biology* **80**, 377 (1997).

———. "Information in Bits and Bytes: Reply to Lifson's Review of 'Information Theory and Molecular Biology'" *BioEssays* **17**, 85 (1995).

———. "Self Organization Origin of Life Scenarios and Information Theory," *Journal of Theoretical Biology* **91**, 14 (1981).

INDEX

ABOUT THE AUTHOR

Dean L. Overman is a senior partner in the Washington, D.C. office of Winston & Strawn, a large international law firm. While practicing in the area of international law, he taught a secured financing course as a member of the faculty of the University of Virginia Law School and also served as a Visiting Scholar and Officer of Harvard University. He was a White House Fellow and served as Special Assistant to Vice President Nelson Rockefeller and as Associate Director of the White House Domestic Council for President Ford. He is the co-author of several law books, the author of many law review articles on banking, commercial, corporate, tax and securities laws, the author of a book on effective writing, and an Adjunct Fellow at the Center for Strategic and International Studies. He received his Juris Doctor from the University of California at Berkeley (Boalt Hall) and did graduate work at the University of Chicago and Princeton Theological Seminary. He is a member of the Triple Nine Society and the International Society for Philosophical Enquiry.